SELECTED SOLUTIONS MANUAL

Karen C. Timberlake

Los Angeles Valley College

CHEMISTRY

An Introduction to General, Organic,
and Biological Chemistry

ELEVENTH EDITION

Prentice Hall

Boston Columbus Indianapolis New York San Francisco Upper Saddle River
Amsterdam Cape Town Dubai London Madrid Milan Munich Paris Montréal Toronto
Delhi Mexico City São Paulo Sydney Hong Kong Seoul Singapore Taipei Tokyo

Editor in Chief: Adam Jaworski
Marketing Manager: Erin Gardner
Associate Editor: Jessica Neumann
Editorial Assistant: Catherine Martinez
Marketing Assistant: Nicola Houston
Managing Editor, Chemistry and Geosciences: Gina M. Cheselka
Project Manager, Science: Ed Thomas
Supplement Cover Designer: Paul Gourhan
Operations Specialist: Maura Zaldivar
Cover Photo Credit: Pink flamingos sleeping. Corbis photo.

© 2012, 2009, 2006

Pearson Education, Inc.

Pearson Prentice Hall

Upper Saddle River, NJ 07458

Printed in the United States of America

10 9 8 7 6 5 4 3 2 1

ISBN-13:   978-0-321-76521-5
ISBN-10:   0-321-76521-4

Prentice Hall
is an imprint of

www.pearsonhighered.com

Contents

1

Chemistry and Measurements

1.1 Many chemicals are listed on a vitamin bottle such as vitamin A, vitamin B_3, vitamin B_{12}, vitamin C, folic acid, etc.

1.3 No. All of the ingredients listed are chemicals.

1.5 Among the things you might do to help yourself succeed in chemistry: attend lecture regularly, review the *Learning Goals*, keep a problem notebook, read the text actively, read the chapter before lecture, join a study group, use your instructor's office hours, etc.

1.7 Ways you can enhance your learning of chemistry:
   a. Form a study group.
   c. Visit the professor during office hours.
   e. Become an active learner.
   f. Work the exercises in the *Study Guide*.

1.9 In the United States, **a.** body mass is measured in pounds (lb), **b.** height in feet and inches, **c.** amount of gasoline in gallons, and **d.** temperature in degrees Fahrenheit (°F). In Mexico, **a.** body mass is measured in kilograms (kg), **b.** height in meters, **c.** amount of gasoline in liters, and **d.** temperature in degrees Celsius (°C).

1.11 **a.** The unit is a meter, which is a unit of length.
   b. The unit is a gram, which is a unit of mass.
   c. The unit is a milliliter, which is a unit of volume.
   d. The unit is a second, which is a unit of time.
   e. The unit is a degree Celsius, which is a unit of temperature.

1.13 **a.** Move the decimal point four places to the left to give 5.5×10^4 m.

   b. Move the decimal point two places to the left to give 4.8×10^2 g.

   c. Move the decimal point six places to the right to give 5×10^{-6} cm.

   d. Move the decimal point four places to the right to give 1.4×10^{-4} s.

   e. Move the decimal point three places to the right to give 7.2×10^{-3} L.

   f. Move the decimal point five places to the left to give 6.7×10^5 kg.

1.15 **a.** The value 7.2×10^3 cm, which is also 72×10^2 cm, is larger than 8.2×10^2 cm.

   b. The value 3.2×10^{-2} kg, which is also 320×10^{-4} kg, is larger than 4.5×10^{-4} kg.

   c. The value 1×10^4 L or 10 000 L is larger than 1×10^{-4} L or 0.0001 L.

   d. The value 6.8×10^{-2} m or 0.068 m is larger than 0.000 52 m.

1.17   Measured numbers are obtained using some type of measuring device. Exact numbers are numbers obtained by counting items or using a definition that compares two units in the same measuring system.

    a.   The value 155 lb is a measured number.

    b.   The value 2 tablets is obtained by counting, making it an exact number.

    c.   The values in the definition 1 kg = 1000 g are exact numbers.

    d.   The value 1720 km is a measured number.

1.19   Measured numbers are obtained using some type of measuring device. Exact numbers are numbers obtained by counting items or using a definition that compares two units in the same measuring system.

    a.   3 hamburgers is a counted/exact number, whereas the value 6 oz of hamburger meat is obtained by measurement.

    b.   Neither are measured numbers; both 1 table and 4 chairs are counted/exact numbers.

    c.   Both 0.75 lb of grapes and 350 g of butter are obtained by measurement.

    d.   Neither are measured numbers; the values in a definition are exact numbers.

1.21   **a.**   Zeros at the beginning of a decimal number are *not significant*.

    b.   Zeros between nonzero digits are *significant*.

    c.   Zeros at the end of a decimal number are *significant*.

    d.   Zeros in the coefficient of a number written in scientific notation are *significant*.

    e.   Zeros used as placeholders in a large number without a decimal point are *not significant*.

1.23   **a.**   All five numbers are significant figures (5 SFs).

    b.   Only the two nonzero numbers are significant (2 SFs); the preceding zeros are placeholders.

    c.   Only the two nonzero numbers are significant (2 SFs); the zeros that follow are placeholders.

    d.   All three numbers in the coefficient of a number written in scientific notation are significant (3 SFs).

    e.   All four numbers to the right of the decimal point, including the last zero, in a decimal number are significant (4 SFs).

    f.   All three numbers including the zeros at the end of a decimal number are significant (3 SFs).

1.25   Both measurements in part **c** have two significant figures.

1.27   A calculator often gives more digits than the number of significant figures allowed in the answer.

1.29   **a.**   1.85 kg; the last digit is dropped since it is 4 or less.

    b.   88.2 L; since the fourth digit is 4 or less, the last three digits are dropped.

    c.   0.004 74 cm; since the fourth significant digit (the first digit to be dropped) is 5 or greater, the last retained digit is increased by 1 when the last four digits are dropped.

    d.   8810 m; since the fourth significant digit (the first digit to be dropped) is 5 or greater, the last retained digit is increased by 1 when the last digit is dropped (a nonsignificant zero is added at the end as a placeholder).

    e. 1.83×10^5 s; since the fourth digit is 4 or less, the last digit is dropped. The $\times 10^5$ is retained so that the magnitude of the answer is not changed.

1.31   **a.**   As written, 5080 L only has three significant figures; this can be shown in scientific notation as 5.08×10^3 L.

    b.   As written, 37 400 g only has three significant figures; this can be shown in scientific notation as 3.74×10^4 g.

    c.   To round off 104 720 m to three significant figures, drop the final digits 720 and increase the last retained digit by 1 and add placeholder zeros to give 105 000 m or 1.05×10^5 m.

    d.   To round off 0.000 250 82 s to three significant figures, drop the final digits 82 and increase the last retained digit by 1 to give 0.000 251 s or 2.51×10^{-4} s.

1.33 **a.** $45.7 \times 0.034 = 1.6$ Two significant figures are allowed since 0.034 has 2 SFs.

b. $0.002\,78 \times 5 = 0.01$ One significant figure is allowed since 5 has 1 SF.

c. $\dfrac{34.56}{1.25} = 27.6$ Three significant figures are allowed since 1.25 has 3 SFs.

d. $\dfrac{(0.2465)(25)}{1.78} = 3.5$ Two significant figures are allowed since 25 has 2 SFs.

1.35 **a.** $45.48 \text{ cm} + 8.057 \text{ cm} = 53.54 \text{ cm}$ Two decimal places are allowed since 45.48 cm has two decimal places.

b. $23.45 \text{ g} + 104.1 \text{ g} + 0.025 \text{ g} = 127.6 \text{ g}$ One decimal place is allowed since 104.1 g has one decimal place.

c. $145.675 \text{ mL} - 24.2 \text{ mL} = 121.5 \text{ mL}$ One decimal place is allowed since 24.2 mL has one decimal place.

d. $1.08 \text{ L} - 0.585 \text{ L} = 0.50 \text{ L}$ Two decimal places are allowed since 1.08 L has two decimal places.

1.37 The km/h markings indicate how many kilometers (how much distance) will be traversed in 1 hour's time if the speed is held constant. The mph (mi/h) markings indicate the same distance traversed but measured in miles during the 1 hour of travel.

1.39 Because the prefix *kilo* means to multiply by 1000, 1 kg is the same mass as 1000 g.

1.41 **a.** mg
b. dL
c. km
d. pg
e. μL
f. ns

1.43 **a.** 0.01

b. 1 000 000 000 000 (or 1×10^{12})

c. 0.001 (or 1×10^{-3})

d. 0.1

e. 1 000 000 (or 1×10^{6})

f. 0.000 000 000 001 (or 1×10^{-12})

1.45 **a.** $1 \text{ m} = 100 \text{ cm}$

b. $1 \text{ m} = 1 \times 10^{9} \text{ nm}$

c. $1 \text{ mm} = 0.001 \text{ m}$

d. $1 \text{ L} = 1000 \text{ mL}$

1.47 **a.** kilogram, since 10^{3} g is greater than 10^{-3} g.

b. milliliter, since 10^{-3} L is greater than 10^{-6} L.

c. km, since 10^{3} m is greater than 10^{0} m.

d. kL, since 10^{3} L is greater than 10^{-1} L.

e. nanometer, since 10^{-9} m is greater than 10^{-12} m.

1.49   A conversion factor can be inverted to give a second conversion factor: $\dfrac{1\ m}{100\ cm}$ and $\dfrac{100\ cm}{1\ m}$

1.51   **a.**   1 m = 100 cm; $\dfrac{1\ m}{100\ cm}$ and $\dfrac{100\ cm}{1\ m}$

      b.   1 g = 1000 mg; $\dfrac{1\ g}{1000\ mg}$ and $\dfrac{1000\ mg}{1\ g}$

      c.   1 L = 1000 mL; $\dfrac{1\ L}{1000\ mL}$ and $\dfrac{1000\ mL}{1\ L}$

      d.   1 dL = 100 mL; $\dfrac{1\ dL}{100\ mL}$ and $\dfrac{100\ mL}{1\ dL}$

      e.   1 week = 7 days; $\dfrac{1\ week}{7\ days}$ and $\dfrac{7\ days}{1\ week}$

1.53   **a.**   1 s = 3.5 m; $\dfrac{3.5\ m}{1\ s}$ and $\dfrac{1\ s}{3.5\ m}$

      b.   1 day = 3500 mg of potassium; $\dfrac{3500\ mg\ potassium}{1\ day}$ and $\dfrac{1\ day}{3500\ mg\ potassium}$

      c.   1.0 gal of gasoline = 46.0 km; $\dfrac{46.0\ km}{1.0\ gal\ gasoline}$ and $\dfrac{1.0\ gal\ gasoline}{46.0\ km}$

      d.   1 tablet = 50 mg of Atenolol; $\dfrac{50\ mg\ Atenolol}{1\ tablet}$ and $\dfrac{1\ tablet}{50\ mg\ Atenolol}$

      e.   1 kg of plums = 29 μg of pesticide; $\dfrac{29\ \mu g\ pesticide}{1\ kg\ plums}$ and $\dfrac{1\ kg\ plums}{29\ \mu g\ pesticide}$

      f.   1 tablet = 81 mg of aspirin; $\dfrac{81\ mg\ aspirin}{1\ tablet}$ and $\dfrac{1\ tablet}{81\ mg\ aspirin}$

1.55   When using a conversion factor, you are trying to cancel existing units and arrive at a new (needed) unit. The conversion factor must be properly oriented so that the unit in the denominator cancels the preceding unit in the numerator.

1.57   **a.**   **Given** 175 cm   **Need** meters

        Plan   cm $\rightarrow$ m $\dfrac{1\ m}{100\ cm}$

        Set Up   175 $\cancel{cm} \times \dfrac{1\ m}{100\ \cancel{cm}} = 1.75$ m (3 SFs)

      b.   **Given** 5500 mL   **Need** liters

        Plan   mL $\rightarrow$ L $\dfrac{1\ L}{1000\ mL}$

        Set Up   5500 $\cancel{mL} \times \dfrac{1\ L}{1000\ \cancel{mL}} = 5.5$ L (2 SFs)

c. **Given** 0.0018 kg   **Need** grams

Plan kg → g $\dfrac{1000 \text{ g}}{1 \text{ kg}}$

Set Up 0.0018 kg $\times \dfrac{1000 \text{ g}}{1 \text{ kg}} = 1.8$ g (2 SFs)

1.59   a. **Given** 0.500 qt   **Need** milliliters

Plan qt → mL $\dfrac{946 \text{ mL}}{1 \text{ qt}}$

Set Up 0.500 qt $\times \dfrac{946 \text{ mL}}{1 \text{ qt}} = 473$ mL (3 SFs)

b. **Given** 145 lb   **Need** kilograms

Plan lb → kg $\dfrac{1 \text{ kg}}{2.20 \text{ lb}}$

Set Up 145 lb $\times \dfrac{1 \text{ kg}}{2.20 \text{ lb}} = 65.9$ kg (3 SFs)

c. **Given** 74 kg body mass, 15% body fat   **Need** pounds of body fat

Plan kg of body mass → kg of body fat → lb of body fat

$\dfrac{15 \text{ kg body fat}}{100 \text{ kg body mass}}$ $\dfrac{2.20 \text{ lb body fat}}{1 \text{ kg body fat}}$

Set Up 74 kg body mass $\times \dfrac{15 \text{ kg body fat}}{100 \text{ kg body mass}} \times \dfrac{2.20 \text{ lb body fat}}{1 \text{ kg body fat}} = 24$ lb of body fat (2 SFs)

d. **Given** 10.0 oz of fertilizer   **Need** grams of nitrogen

Plan oz of fertilizer → lb of fertilizer → g of fertilizer → g of nitrogen
(percent equality: 100 g of fertilizer = 15 g of nitrogen) $\dfrac{1 \text{ lb}}{16 \text{ oz}}$ $\dfrac{454 \text{ g}}{1 \text{ lb}}$ $\dfrac{15 \text{ g nitrogen}}{100 \text{ g fertilizer}}$

Set Up 10.0 oz fertilizer $\times \dfrac{1 \text{ lb}}{16 \text{ oz}} \times \dfrac{454 \text{ g}}{1 \text{ lb}} \times \dfrac{15 \text{ g nitrogen}}{100 \text{ g fertilizer}} = 43$ g of nitrogen (2 SFs)

e. **Given** 5.0 kg of pecans   **Need** pounds of chocolate bars

Plan kg of pecans → kg of bars → lb of bars
(percent equality: 100 kg of bars = 22.0 kg of pecans) $\dfrac{100 \text{ kg choc. bars}}{22.0 \text{ kg pecans}}$ $\dfrac{2.20 \text{ lb}}{1 \text{ kg}}$

Set Up 5.0 kg pecans $\times \dfrac{100 \text{ kg choc. bars}}{22.0 \text{ kg pecans}} \times \dfrac{2.20 \text{ lb}}{1 \text{ kg}} = 50$ lb of chocolate bars (2 SFs)

1.61   **a.**   **Given** 250 L of water   **Need** gallons of water

  Plan $L \rightarrow qt \rightarrow gal$ $\dfrac{1.06\ qt}{1\ L}$ $\dfrac{1\ gal}{4\ qt}$

  Set Up $250\ \cancel{L} \times \dfrac{1.06\ \cancel{qt}}{1\ \cancel{L}} \times \dfrac{1\ gal}{4\ \cancel{qt}} = 66\ gal\ (2\ SFs)$

  b.   **Given** 0.024 g of sulfa drug, 8-mg tablets   **Need** number of tablets

  Plan g of sulfa drug $\rightarrow$ mg of sulfa drug $\rightarrow$ number of tablets $\dfrac{1000\ mg}{1\ g}$ $\dfrac{1\ tablet}{8\ mg\ sulfa\ drug}$

  Set Up $0.24\ \cancel{g\ sulfa\ drug} \times \dfrac{1000\ \cancel{mg}}{1\ \cancel{g}} \times \dfrac{1\ tablet}{8\ \cancel{mg\ sulfa\ drug}} = 3\ tablets\ (1\ SF)$

  c.   **Given** 34-lb child, 115 mg of ampicillin/kg of body mass   **Need** milligrams of ampicillin

  Plan lb of body mass $\rightarrow$ kg of body mass $\rightarrow$ mg of ampicillin $\dfrac{1\ kg}{2.20\ lb}$ $\dfrac{115\ mg\ ampicillin}{1\ kg\ body\ mass}$

  Set Up $34\ \cancel{lb\ body\ mass} \times \dfrac{1\ \cancel{kg\ body\ mass}}{2.20\ \cancel{lb\ body\ mass}} \times \dfrac{115\ mg\ ampicillin}{1\ \cancel{kg\ body\ mass}}$

  $= 1800\ mg\ of\ ampicillin\ (2\ SFs)$

1.63   Because the density of aluminum is 2.70 g/cm^3, silver is 10.5 g/cm^3, and lead is 11.3 g/cm^3, we can identify the unknown metal by calculating its density as follows:

$$Density = \frac{mass\ of\ metal}{volume\ of\ metal} = \frac{217\ g}{19.2\ cm^3} = 11.3\ g/cm^3\ (3\ SFs)$$

∴ the metal is lead.

1.65   Density is the mass of a substance divided by its volume. $Density = \dfrac{mass\ (grams)}{volume\ (mL)}$

The densities of solids and liquids are usually stated in g/mL or g/cm^3, so in some problems the units will need to be converted.

  a. $Density = \dfrac{mass\ (grams)}{volume\ (mL)} = \dfrac{24.0\ g}{20.0\ mL} = 1.20\ g/mL\ (3\ SFs)$

  b.   **Given** 1.65 lb, 170 mL   **Need** density (g/mL)

  Plan $lb \rightarrow g$ then calculate density $\dfrac{454\ g}{1\ lb}$

  Set Up $1.65\ \cancel{lb} \times \dfrac{454\ g}{1\ \cancel{lb}} = 749\ g\ (3\ SFs)$

  ∴ $Density = \dfrac{mass}{volume} = \dfrac{749\ g}{170\ mL} = 4.4\ g/mL\ (2\ SFs)$

c. **Given** 20.0 mL initial volume, 34.5 mL final volume, 45.0 g   **Need** density (g/mL)

Plan calculate volume by difference, then calculate density

Set Up volume of gem: 34.5 mL total − 20.0 mL water = 14.5 mL

$$\therefore \text{Density} = \frac{\text{mass}}{\text{volume}} = \frac{45.0 \text{ g}}{14.5 \text{ ml}} = 3.10 \text{ g/mL (3 SFs)}$$

d. **Given** 514.1 g, 114 cm^3 **Need** density (g/mL)

Plan convert volume cm^3 → mL   then calculate the density

Set Up $114 \text{ cm}^3 \times \dfrac{1 \text{ mL}}{1 \text{ cm}^3} = 114 \text{ mL}$

$$\therefore \text{Density} = \frac{\text{mass}}{\text{volume}} = \frac{514.1 \text{ g}}{114 \text{ mL}} = 4.51 \text{ g/mL (3 SFs)}$$

1.67 In these problems, the density is used as a conversion factor.
  a. **Given** 150 mL of liquid, density 1.4 g/mL   **Need** grams of liquid

Plan mL → g $\dfrac{1.4 \text{ g}}{1 \text{ mL}}$

Set Up $150 \text{ mL} \times \dfrac{1.4 \text{ g}}{1 \text{ mL}} = 210 \text{ g of liquid (2 SFs)}$

  b. **Given** 0.500 L of glucose solution, density 1.15 g/mL   **Need** grams of solution

Plan L → mL → g $\dfrac{1000 \text{ mL}}{1 \text{ L}}$ $\dfrac{1.15 \text{ g}}{1 \text{ mL solution}}$

Set Up $0.500 \text{ L solution} \times \dfrac{1000 \text{ mL}}{1 \text{ L}} \times \dfrac{1.15 \text{ g}}{1 \text{ mL solution}} = 575 \text{ g of solution (3 SFs)}$

  c. **Given** 225 mL of bronze   **Need** ounces of bronze

Plan mL → g → lb → oz $\dfrac{7.8 \text{ g}}{1 \text{ mL}}$ $\dfrac{1 \text{ lb}}{454 \text{ g}}$ $\dfrac{16 \text{ oz}}{1 \text{ lb}}$

Set Up $225 \text{ mL} \times \dfrac{7.8 \text{ g}}{1 \text{ mL}} \times \dfrac{1 \text{ lb}}{454 \text{ g}} \times \dfrac{16 \text{ oz}}{1 \text{ lb}} = 62 \text{ oz of bronze (2 SFs)}$

1.69 A successful study plan would include:
  b. Working the *Sample Problems* as you go through a chapter.
  c. Going to your professor's office hours.

1.71 Both measurements in part **c** have two significant figures, and both measurements in part **d** have four significant figures.

1.73 a. The number of legs is a counted number; it is exact.
  b. The height is measured with a ruler or tape measure; it is a measured number.
  c. The number of chairs is a counted number; it is exact.
  d. The area is measured with a ruler or tape measure; it is a measured number.

1.75   **a.**   length = 6.96 cm; width = 4.75 cm (Each answer may vary in the estimated digit.)

   b.   length = 69.6 mm; width = 47.5 mm

   c.   There are three significant figures in the length measurement.

   d.   There are three significant figures in the width measurement.

   e.   Area = length × width = 6.96 cm × 4.75 cm = 33.1 cm^2

   f.   Since there are three significant figures in the width and length measurements, there are three significant figures in the calculated area.

1.77   **Given** 18.5 mL initial volume, 23.1 mL final volume, 8.24 g mass   **Need** density (g/mL)

   Plan calculate volume by difference, then calculate density

   Set Up The volume of the object is 23.1 mL − 18.5 mL = 4.6 mL

$$\therefore \text{Density} = \frac{\text{mass}}{\text{volume}} = \frac{8.24 \text{ g}}{4.6 \text{ mL}} = 1.8 \text{ g/mL (2 SFs)}$$

1.79   **a.**   To round off 0.000 012 58 L to three significant figures, drop the final digit 8 and increase the last retained digit by 1 to give 0.000 012 6 L or 1.26×10^{-5} L.

   b.   To round off 3.528×10^2 kg to three significant figures, drop the final digit 8 and increase the last retained digit by 1 to give 3.53×10^2 kg.

   c.   To round off 125 111 m to three significant figures, drop the final digits 111 and add three zeros as placeholders to give 125 000 m (or 1.25×10^5 m).

   d.   To express 58.703 g to three significant figures, drop the final digits 03 to give 58.7 g.

   e.   To express 3×10^{-3} s to three significant figures, add two significant zeros to give 3.00×10^{-3} s.

   f.   To round off 0.010 826 g to three significant figures, drop the final digits 26 to give 0.0108 g or 1.08×10^{-2} g.

1.81   This problem requires several conversion factors. When you write out the unit plan, be sure you know a conversion factor you can use for each step. A possible unit plan follows:

   Given 7500 ft   **Need** minutes

   Plan ft → in. → cm → m → min $\dfrac{12 \text{ in.}}{1 \text{ ft}}$ $\dfrac{2.54 \text{ cm}}{1 \text{ in.}}$ $\dfrac{1 \text{ m}}{100 \text{ cm}}$ $\dfrac{1 \text{ min}}{55.0 \text{ m}}$

   Set Up $7500 \text{ ft} \times \dfrac{12 \text{ in.}}{1 \text{ ft}} \times \dfrac{2.54 \text{ cm}}{1 \text{ in.}} \times \dfrac{1 \text{ m}}{100 \text{ cm}} \times \dfrac{1 \text{ min}}{55.0 \text{ m}} = 42 \text{ min (2 SFs)}$

1.83   **Given** 1.75 Euros/kg of grapes, \$1.36/Euro   **Need** cost in dollars per pound

   Plan Euros/kg → Euros/lb → \$/lb $\dfrac{1.75 \text{ Euros}}{1 \text{ kg grapes}}$ $\dfrac{1 \text{ kg}}{2.20 \text{ lb}}$ $\dfrac{\$1.36}{1 \text{ Euro}}$

   Set Up $\dfrac{1.75 \text{ Euros}}{1 \text{ kg grapes}} \times \dfrac{1 \text{ kg grapes}}{2.20 \text{ lb}} \times \dfrac{\$1.36}{1 \text{ Euro}} = \$1.08 / \text{lb of grapes (3 SFs)}$

1.85 **Given** 4.0 lb of onions   **Need** number of onions

Plan lb → g → number of onions $\dfrac{454 \text{ g}}{1 \text{ lb}}$ $\dfrac{1 \text{ onion}}{115 \text{ g}}$

Set Up $4.0 \text{ lb onions} \times \dfrac{454 \text{ g}}{1 \text{ lb}} \times \dfrac{1 \text{ onion}}{115 \text{ g}} = 16 \text{ onions (2 SFs)}$

1.87 **a.** **Given** 8.0 oz   **Need** number of crackers

Plan oz → number of crackers $\dfrac{6 \text{ crackers}}{0.50 \text{ oz}}$

Set Up $8.0 \text{ oz} \times \dfrac{6 \text{ crackers}}{0.5 \text{ oz}} = 96 \text{ crackers (2 SFs)}$

b. **Given** 10 crackers   **Need** ounces of fat

Plan number of crackers → servings → g of fat → lb of fat → oz of fat

$\dfrac{1 \text{ serving}}{6 \text{ crackers}}$ $\dfrac{4 \text{ g fat}}{1 \text{ serving}}$ $\dfrac{1 \text{ lb}}{454 \text{ g}}$ $\dfrac{16 \text{ oz}}{1 \text{ lb}}$

Set Up $10 \text{ crackers} \times \dfrac{1 \text{ serving}}{6 \text{ crackers}} \times \dfrac{4 \text{ g fat}}{1 \text{ serving}} \times \dfrac{1 \text{ lb}}{454 \text{ g}} \times \dfrac{16 \text{ oz}}{1 \text{ lb}} = 0.2 \text{ oz of fat (1 SF)}$

c. **Given** 50 boxes   **Need** grams of sodium

Plan boxes → oz → servings → mg of sodium → g of sodium

$\dfrac{8.0 \text{ oz}}{1 \text{ box}}$ $\dfrac{1 \text{ serving}}{0.50 \text{ oz}}$ $\dfrac{140 \text{ mg sodium}}{1 \text{ serving}}$ $\dfrac{1 \text{ g}}{1000 \text{ mg}}$

Set Up $50 \text{ boxes} \times \dfrac{8.0 \text{ oz}}{1 \text{ box}} \times \dfrac{1 \text{ serving}}{0.50 \text{ oz}} \times \dfrac{140 \text{ mg sodium}}{1 \text{ serving}} \times \dfrac{1 \text{ g sodium}}{1000 \text{ mg sodium}}$

$= 110 \text{ g of sodium (2 SFs)}$

1.89 **Given** 10 days, 4 tablets/day, 250-mg tablets   **Need** ounces of amoxicillin

Plan days → tablets → mg of amoxicillin → g → lb → oz of amoxicillin

$\dfrac{4 \text{ tablets}}{1 \text{ day}}$ $\dfrac{250 \text{ mg amoxicillin}}{1 \text{ tablet}}$ $\dfrac{1 \text{ g}}{1000 \text{ mg}}$ $\dfrac{1 \text{ lb}}{454 \text{ g}}$ $\dfrac{16 \text{ oz}}{1 \text{ lb}}$

Set Up $10 \text{ days} \times \dfrac{4 \text{ tablets}}{1 \text{ day}} \times \dfrac{250 \text{ mg amoxicillin}}{1 \text{ tablet}} \times \dfrac{1 \text{ g}}{1000 \text{ mg}} \times \dfrac{1 \text{ lb}}{454 \text{ g}} \times \dfrac{16 \text{ oz}}{1 \text{ lb}}$

$= 0.35 \text{ oz of amoxicillin (2 SFs)}$

1.91   **Given** 1.85 g/L   **Need** mg/dL

Plan g/L → mg/L → mg/dL $\dfrac{1000 \text{ mg}}{1 \text{ g}}$ $\dfrac{1 \text{ L}}{10 \text{ dL}}$

Set Up This problem involves two unit conversions. Convert g to mg in the numerator, and convert L in the denominator to dL.

$$\frac{1.85 \text{ g}}{1 \text{ L}} \times \frac{1000 \text{ mg}}{1 \text{ g}} \times \frac{1 \text{ L}}{10 \text{ dL}} = 185 \text{ mg/dL (3 SFs)}$$

1.93   **Given** 215 mL initial, 285 mL final volume   **Need** grams of lead

Plan calculate the volume by difference and   mL → g $\dfrac{11.3 \text{ g}}{1 \text{ mL}}$

Set Up The difference between the initial volume of the water and its volume with the lead object will give us the volume of the lead object: 285 mL total − 215 mL water = 70. mL of lead, then

$$70. \text{ mL lead} \times \frac{11.3 \text{ g lead}}{1 \text{ mL lead}} = 790 \text{ g of lead (2 SFs)}$$

1.95   **Given** 1.50 L of gasoline   **Need** cm^3 of olive oil

Plan L of gasoline → mL of gasoline → g of gasoline → g of olive oil → cm^3 of olive oil

(equality from question: 1 g olive oil = 1 g of gasoline)

$$\frac{1000 \text{ mL gasoline}}{1 \text{ L gasoline}} \quad \frac{0.74 \text{ g gasoline}}{1 \text{ mL gasoline}} \quad \frac{1 \, cm^3 \text{ olive oil}}{0.92 \text{ g olive oil}}$$

Set Up $1.50 \text{ L gasoline} \times \dfrac{1000 \text{ mL}}{1 \text{ L}} \times \dfrac{0.74 \text{ g gasoline}}{1 \text{ mL gasoline}} \times \dfrac{1 \text{ g olive oil}}{1 \text{ g gasoline}} \times \dfrac{1 \, cm^3 \text{ olive oil}}{0.92 \text{ g olive oil}}$

$$= 1200 \, cm^3 \ (1.2 \times 10^3 \, cm^3) \text{ of olive oil (2 SFs)}$$

1.97   Because the balance can measure mass to 0.001 g, the mass should be reported to 0.001 g. You should record the mass of the object as 31.075 g.

1.99   **Given** 3.0 h   **Need** gallons of gasoline

Plan h → mi → km → L → qt → gal $\dfrac{55 \text{ mi}}{1 \text{ h}}$ $\dfrac{1 \text{ km}}{0.621 \text{ mi}}$ $\dfrac{1 \text{ L}}{11 \text{ km}}$ $\dfrac{1.06 \text{ qt}}{1 \text{ L}}$ $\dfrac{1 \text{ gal}}{4 \text{ qt}}$

Set Up $3.0 \text{ h} \times \dfrac{55 \text{ mi}}{\text{h}} \times \dfrac{1 \text{ km}}{0.621 \text{ mi}} \times \dfrac{1 \text{ L}}{11 \text{ km}} \times \dfrac{1.06 \text{ qt}}{1 \text{ L}} \times \dfrac{1 \text{ gal}}{4 \text{ qt}}$

$$= 6.4 \text{ gal of gasoline (2 SFs)}$$

1.101   **Given** 1.50 g of silicon, 3.00 in. diameter   **Need** thickness (mm)

Plan $g \rightarrow cm^3$ $\dfrac{1\ cm^3}{2.33\ g\ silicon}$

and $d(\text{in.}) \rightarrow r(\text{in.}) \rightarrow r(\text{cm})$ $r = \dfrac{d}{2}\ \dfrac{2.54\ cm}{1\ in.}$

then rearrange Volume equation for thickness (cm) $\rightarrow$ thickness (mm)

$V = \pi r^2 h \rightarrow h = \dfrac{V}{\pi r^2}\ \dfrac{10\ mm}{1\ cm}$

Set Up Volume of wafer: $1.50\ g \times \dfrac{1\ cm^3}{2.33\ g} = 0.644\ cm^3$ (3 SFs)

Radius of wafer: $\dfrac{3.00\ in.}{2} \times \dfrac{2.54\ cm}{1\ in.} = 3.81\ cm$ (3 SFs)

$\therefore h = \dfrac{V}{\pi r^2} = \dfrac{0.644\ cm^3}{\pi (3.81\ cm)^2} = 0.0141\ cm \times \dfrac{10\ mm}{1\ cm} = 0.141\ mm$ (3 SFs)

1.103 **a.** **Given** 180 lb of body mass   **Need** cups of coffee

Plan lb of body mass $\rightarrow$ kg of body mass $\rightarrow$ mg of caffeine $\rightarrow$ fl oz of coffee $\rightarrow$ cups of coffee

$\dfrac{1\ kg\ body\ mass}{2.20\ lb\ body\ mass}$ $\dfrac{192\ mg\ caffeine}{1\ kg\ body\ mass}$ $\dfrac{6\ fl\ oz\ coffee}{100.\ mg\ caffeine}$ $\dfrac{1\ cup\ coffee}{12\ fl\ oz\ coffee}$

Set Up $180\ lb\ body\ mass \times \dfrac{1\ kg\ body\ mass}{2.20\ lb\ body\ mass} \times \dfrac{192\ mg\ caffeine}{1\ kg\ body\ mass}$

$= 1.571 \times 10^4$ mg of caffeine (2 SFs allowed)

$1.571 \times 10^4\ mg\ caffeine \times \dfrac{6\ fl\ oz\ coffee}{100\ mg\ caffeine} \times \dfrac{1\ cup\ coffee}{12\ fl\ oz\ coffee}$

$= 79$ cups of coffee (2 SFs)

b. **Given** mg of caffeine from part **a**   **Need** cans of cola

Plan mg of caffeine $\rightarrow$ cans of cola $\dfrac{1\ can\ cola}{50.\ mg\ caffeine}$

Set Up $1.571 \times 10^4\ mg\ caffeine \times \dfrac{1\ can\ cola}{50.\ mg\ caffeine} = 310$ cans of cola (2 SFs)

c. **Given** mg of caffeine from part **a**   **Need** number of No-Doz tablets

Plan mg of caffeine $\rightarrow$ No-Doz tablets $\dfrac{1\ No\text{-}Doz\ tablet}{100\ mg\ caffeine}$

Set Up $1.571 \times 10^4\ mg\ caffeine \times \dfrac{1\ No\text{-}Doz\ tablet}{100\ mg\ caffeine} = 160$ No-Doz tablets (2 SFs)

1.105   **a.**   **Given** 65 kg of body mass   **Need** pounds of fat

Plan   kg of body mass → kg of fat → lb of fat

(percent equality: 100 kg of body mass = 3.0 kg of fat) $\dfrac{3.0 \text{ kg fat}}{100 \text{ kg body mass}}$ $\dfrac{2.20 \text{ lb fat}}{1 \text{ kg fat}}$

Set Up $65 \text{ kg body mass} \times \dfrac{3.0 \text{ kg fat}}{100 \text{ kg body mass}} \times \dfrac{2.20 \text{ lb fat}}{1 \text{ kg fat}} = 4.3$ lb of fat (2 SFs)

b.   **Given** 3.0 L of fat   **Need** pounds of fat

Plan   L → mL → g → lb $\dfrac{1000 \text{ mL}}{1 \text{ L}}$ $\dfrac{0.94 \text{ g}}{1 \text{ mL}}$ $\dfrac{1 \text{ lb}}{454 \text{ g}}$

Set Up $3.0 \text{ L} \times \dfrac{1000 \text{ mL}}{1 \text{ L}} \times \dfrac{0.94 \text{ g}}{1 \text{ mL}} \times \dfrac{1 \text{ lb}}{454 \text{ g}} = 6.2$ lb of fat (2 SFs)

1.107   The liquid with the highest density will be at the bottom of the cylinder; the liquid with the lowest density will be at the top of the cylinder:

A is vegetable oil (D = 0.92 g/mL), B is water (D = 1.00 g/mL), C is mercury (D = 13.6 g/mL)

2
Matter and Energy

2.1   A *pure substance* is matter that has a fixed or definite composition. Elements and compounds are pure substances. In a *mixture*, two or more substances are physically mixed but not chemically combined.

    a. Baking soda is composed of one type of matter ($NaHCO_3$), which makes it a pure substance.

    b. A blueberry muffin is composed of several substances mixed together, which makes it a mixture.

    c. Ice is composed of one type of matter (H_2O molecules), which makes it a pure substance.

    d. Zinc is composed of one type of matter (Zn atoms), which makes it a pure substance.

    e. Trimix is a physical mixture of oxygen, nitrogen, and helium gases, which makes it a mixture.

2.3   *Elements* are the simplest type of pure substance, containing only one type of atom. *Compounds* contain two or more elements chemically combined in a specific proportion.

    a. A silicon chip is an element since it contains only one type of atom (Si).

    b. Hydrogen peroxide (H_2O_2) is a compound since it contains two elements (H, O) chemically combined.

    c. Oxygen (O_2) is an element since it contains only one type of atom (O).

    d. Rust (Fe_2O_3) is a compound since it contains two elements (Fe, O) chemically combined.

    e. Methane (CH_4) in natural gas is a compound since it contains two elements (C, H) chemically combined.

2.5   A *homogeneous mixture* has a uniform composition; a *heterogeneous mixture* does not have a uniform composition throughout the mixture.

    a. Vegetable soup is a heterogeneous mixture since it has chunks of vegetables.

    b. Sea water is a homogeneous mixture since it has a uniform composition.

    c. Tea is a homogeneous mixture since it has a uniform composition.

    d. Tea with ice and lemon slices is a heterogeneous mixture since it has chunks of ice and lemon.

    e. Fruit salad is a heterogeneous mixture since it has chunks of fruit.

2.7   **a.** A gas has no definite volume or shape.

    b. In a gas, the particles do not interact with each other.

    c. In a solid, the particles are held in a rigid structure.

2.9   A *physical property* is a characteristic of the substance such as color, shape, odor, luster, size, melting point, and density. A *chemical property* is a characteristic that indicates the ability of a substance to change into a new substance.

    a. Color and physical state are physical properties.

    b. The ability to react with oxygen is a chemical property.

    c. The freezing point of a substance is a physical property.

    d. Milk souring describes chemical reactions and so is a chemical property.

    e. Burning butane gas in oxygen forms new substances, which makes it a chemical property.

2.11   A *physical change* describes a change in a physical property that retains the identity of the substance. A *chemical change* occurs when the atoms of the initial substances rearrange to form new substances.

    a. Water vapor condensing is a physical change, since the physical form of the water changes, but not the substance.

    b. Cesium metal reacting is a chemical change, since new substances form.

    c. Gold melting is a physical change, since the physical state changes, but not the substance.

    d. Cutting a puzzle is a physical change, since the size and shape change, but not the substance.

    e. Dissolving sugar is a physical change, since the size of the sugar particles changes, but no new substances are formed.

2.13   **a.** The high reactivity of fluorine is a chemical property, since new substances will be formed.

    b. The physical state of fluorine is a physical property.

    c. The color of fluorine is a physical property.

    d. The reactivity of fluorine with hydrogen is a chemical property, since new substances will be formed.

    e. The melting point of fluorine is a physical property.

2.15   When the roller-coaster car is at the top of the ramp, it has its maximum potential energy. As it descends, potential energy is converted to kinetic energy. At the bottom, all of its energy is kinetic.

2.17   **a.** Potential energy is stored in the water at the top of the waterfall.

    b. Kinetic energy is displayed as the kicked ball moves.

    c. Potential energy is stored in the chemical bonds in the coal.

    d. Potential energy is stored when the skier is at the top of the hill.

2.19   **a.** **Given** 3.0 h, 270 kJ/h   **Need** joules

       Plan $\text{h} \rightarrow \text{kJ} \rightarrow \text{J}$ $\dfrac{1000 \text{ J}}{1 \text{ kJ}}$

       Set Up $3.0 \text{ h} \times \dfrac{270 \text{ kJ}}{1 \text{ h}} \times \dfrac{1000 \text{ J}}{1 \text{ kJ}} = 8.1 \times 10^5 \text{ J}$ (2 SFs)

    b. **Given** 3.0 h, 270 kJ/h   **Need** kilocalories

       Plan $\text{h} \rightarrow \text{kJ} \rightarrow \text{J} \rightarrow \text{cal} \rightarrow \text{kcal}$ $\dfrac{1000 \text{ J}}{1 \text{ kJ}} \quad \dfrac{1 \text{ cal}}{4.184 \text{ J}} \quad \dfrac{1 \text{ kcal}}{1000 \text{ cal}}$

       Set Up $3.0 \text{ h} \times \dfrac{270 \text{ kJ}}{1 \text{ h}} \times \dfrac{1000 \text{ J}}{1 \text{ kJ}} \times \dfrac{1 \text{ cal}}{4.184 \text{ J}} \times \dfrac{1 \text{ kcal}}{1000 \text{ cal}} = 190 \text{ kcal}$ (2 SFs)

2.21   The Fahrenheit temperature scale is still used in the United States. A normal body temperature is 98.6 °F on this scale. To convert her 99.8 °F temperature to the equivalent reading on the Celsius scale, the following calculation must be performed:

$$T_C = \frac{(T_F - 32)}{1.8} = \frac{(99.8 - 32)}{1.8} = \frac{67.8}{1.8} = 37.7 \text{ °C (3 SFs) (1.8 and 32 are exact)}$$

Because a normal body temperature is 37.0 on the Celsius scale, her temperature of 37.7 °C would be a mild fever.

2.23   To convert Celsius to Fahrenheit: $T_F = 1.8(T_C) + 32$

     To convert Fahrenheit to Celsius: $T_C = \dfrac{(T_F - 32)}{1.8}$ (1.8 and 32 are exact numbers)

     To convert Celsius to Kelvin: $T_K = T_C + 273$

     To convert Kelvin to Celsius: $T_C = T_K - 273$

a. $T_F = 1.8(T_C) + 32 = 1.8(37.0) + 32 = 66.6 + 32 = 98.6 \ ^\circ F$

b. $T_C = \dfrac{(T_F - 32)}{1.8} = \dfrac{(65.3 - 32)}{1.8} = \dfrac{33.3}{1.8} = 18.5 \ ^\circ C$

c. $T_K = T_C + 273 = -27 + 273 = 246 \ K$

d. $T_K = T_C + 273 = 62 + 273 = 335 \ K$

e. $T_C = \dfrac{(T_F - 32)}{1.8} = \dfrac{(114 - 32)}{1.8} = \dfrac{82}{1.8} = 46 \ ^\circ C$

2.25 a. $T_C = \dfrac{(T_F - 32)}{1.8} = \dfrac{(106 - 32)}{1.8} = \dfrac{74}{1.8} = 41 \ ^\circ C$

b. $T_C = \dfrac{(T_F - 32)}{1.8} = \dfrac{(103 - 32)}{1.8} = \dfrac{71}{1.8} = 39 \ ^\circ C$

No, there is no need to phone the doctor. The child's temperature is less than 40.0 °C.

2.27 Copper, which has the lowest specific heat of the samples, would reach the highest temperature.

2.29 a. **Given** $SH_{water} = 1.00 \ cal/g \ ^\circ C$; mass = 8.5 g; $\Delta T = 36 \ ^\circ C - 15 \ ^\circ C = 21 \ ^\circ C$
    Need heat in calories

    Plan Heat = mass $\times \Delta T \times SH$

    Set Up Heat $= 8.5 \ \cancel{g} \times 21 \ \cancel{^\circ C} \times \dfrac{1.00 \ cal}{\cancel{g} \ \cancel{^\circ C}} = 180 \ cal \ (2 \ SFs)$

b. **Given** $SH_{water} = 4.184 \ J/g \ ^\circ C$; mass = 75 g; $\Delta T = 66 \ ^\circ C - 22 \ ^\circ C = 44 \ ^\circ C$
    Need heat in joules

    Plan Heat = mass $\times \Delta T \times SH$

    Set Up Heat $= 75 \ \cancel{g} \times 44 \ \cancel{^\circ C} \times \dfrac{4.184 \ J}{\cancel{g} \ \cancel{^\circ C}} = 14 \ 000 \ J \ (2 \ SFs)$

c. **Given** $SH_{water} = 1.00 \ cal/g \ ^\circ C$; mass = 150 g; $\Delta T = 77 \ ^\circ C - 15 \ ^\circ C = 62 \ ^\circ C$
    Need heat in kilocalories

    Plan Heat = mass $\times \Delta T \times SH$ then cal $\rightarrow$ kcal $\dfrac{1 \ kcal}{1000 \ cal}$

    Set Up Heat $= 150 \ \cancel{g} \times 62 \ \cancel{^\circ C} \times \dfrac{1.00 \ \cancel{cal}}{\cancel{g} \ \cancel{^\circ C}} \times \dfrac{1 \ kcal}{1000 \ \cancel{cal}} = 9.3 \ kcal \ (2 \ SFs)$

d. **Given** $SH_{copper} = 0.385 \ J/g \ ^\circ C$; mass = 175 g; $\Delta T = 188 \ ^\circ C - 28 \ ^\circ C = 160. \ ^\circ C$
    Need heat in kilojoules

    Plan Heat = mass $\times \Delta T \times SH$ then J $\rightarrow$ kJ $\dfrac{1 \ kJ}{1000 \ J}$

    Set Up Heat $= 175 \ \cancel{g} \times 160. \ \cancel{^\circ C} \times \dfrac{0.385 \ \cancel{J}}{\cancel{g} \ \cancel{^\circ C}} \times \dfrac{1 \ kJ}{1000 \ \cancel{J}} = 10.8 \ kJ \ (3 \ SFs)$

2.31 **a.** **Given** $SH_{water} = 4.184$ J/g °C $= 1.00$ cal/g °C; mass $= 25.0$ g; $\Delta T = 25.7$ °C $- 12.5$ °C $= 13.2$ °C

Need heat in joules and calories

Plan Heat $=$ mass $\times \Delta T \times SH$

Set Up Heat $= 25.0 \text{ g} \times 13.2 \text{ °C} \times \dfrac{4.184 \text{ J}}{\text{g °C}} = 1380$ J (3 SFs)

or Heat $= 25.0 \text{ g} \times 13.2 \text{ °C} \times \dfrac{1.00 \text{ cal}}{\text{g °C}} = 330.$ cal (3 SFs)

b. **Given** $SH_{copper} = 0.385$ J/g °C $= 0.0920$ cal/g °C; mass $= 38.0$ g;
$\Delta T = 246$ °C $- 122$ °C $= 124$ °C

Need heat in joules and calories

Plan Heat $=$ mass $\times \Delta T \times SH$

Set Up Heat $= 38.0 \text{ g} \times 124 \text{ °C} \times \dfrac{0.385 \text{ J}}{\text{g °C}} = 1810$ J (3 SFs)

or Heat $= 38.0 \text{ g} \times 124 \text{ °C} \times \dfrac{0.0920 \text{ cal}}{\text{g °C}} = 434$ cal (3 SFs)

c. **Given** $SH_{ethanol} = 2.46$ J/g °C $= 0.588$ cal/g °C; mass $= 15.0$ g;
$\Delta T = -42.0$ °C $- 60.5$ °C $= -102.5$ °C

Need heat in joules and calories

Plan Heat $=$ mass $\times \Delta T \times SH$

Set Up Heat $= 15.0 \text{ g} \times (-102.5 \text{ °C}) \times \dfrac{2.46 \text{ J}}{\text{g °C}} = -3780$ J (3 SFs)

or Heat $= 15.0 \text{ g} \times (-102.5 \text{ °C}) \times \dfrac{0.588 \text{ cal}}{\text{g °C}} = -904$ cal (3 SFs)

d. **Given** $SH_{iron} = 0.452$ J/g °C $= 0.108$ cal/g °C; mass $= 125$ g; $\Delta T = 55$ °C $- 118$ °C $= -63$ °C

Need heat in joules and calories

Plan Heat $=$ mass $\times \Delta T \times SH$

Set Up Heat $= 125 \text{ g} \times (-63 \text{ °C}) \times \dfrac{0.452 \text{ J}}{\text{g °C}} = -3600$ J (2 SFs)

or Heat $= 125 \text{ g} \times (-63 \text{ °C}) \times \dfrac{0.108 \text{ cal}}{\text{g °C}} = -850$ cal (2 SFs)

2.33 **a.** **Given** mass = 505 g of water; $\Delta T = 35.7\ °C - 25.5\ °C = 10.5\ °C$; $SH_{water} = 1.00\ cal/g\ °C$

Need energy in kilocalories

Plan $Heat = mass \times \Delta T \times SH$ then $cal \rightarrow kcal$ $\dfrac{1\ kcal}{1000\ cal}$

Set Up $505\ \not{g} \times 10.5\ \not{°C} \times \dfrac{1.00\ \not{cal}}{\not{g}\ \not{°C}} \times \dfrac{1\ kcal}{1000\ \not{cal}} = 5.30\ kcal$ (3 SFs)

b. **Given** mass = 4980 g of water; $\Delta T = 62.4\ °C - 20.6\ °C = 41.8\ °C$; $SH_{water} = 1.00\ cal/g\ °C$

Need energy in kilocalories

Plan $Heat = mass \times \Delta T \times SH$ then $cal \rightarrow kcal$ $\dfrac{1\ kcal}{1000\ cal}$

Set Up $4980\ \not{g} \times 41.8\ \not{°C} \times \dfrac{1.00\ \not{cal}}{\not{g}\ \not{°C}} \times \dfrac{1\ kcal}{1000\ \not{cal}} = 208\ kcal$ (3 SFs)

2.35 **a.** **Given** 1 cup of orange juice contains 26 g of carbohydrate, 2 g of protein, and no fat
Need total energy in kilojoules

Food Type	Mass	Energy Values		Energy
Carbohydrate	26 $\not{g}$	×	$\dfrac{17\ kJ}{1\ \not{g}}$ =	440 kJ
Protein	2 $\not{g}$	×	$\dfrac{17\ kJ}{1\ \not{g}}$ =	30 kJ
		Total energy content	=	470 kJ (energy results are rounded off to tens place)

b. **Given** one apple provides 72 Cal of energy and contains no fat or protein
Need grams of carbohydrate

Plan $Cal \rightarrow kcal \rightarrow g$ of carbohydrate $\dfrac{1\ kcal}{1\ Cal}$ $\dfrac{1\ g\ of\ carbohydrate}{4\ kcal}$

Set Up $72\ \not{Cal} \times \dfrac{1\ \not{kcal}}{1\ \not{Cal}} \times \dfrac{1\ g\ carbohydrate}{4\ \not{kcal}} = 18\ g$ of carbohydrate (2 SFs)

c. **Given** 1 tablespoon of vegetable oil contains 14 g of fat, and no carbohydrate or protein
Need total energy in kilocalories

Food Type	Mass	Energy Values		Energy
Fat	14 $\not{g}$	×	$\dfrac{9\ kcal}{1\ \not{g}}$ =	130 kcal (rounded off to tens place)

d. **Given** a diet that consists of 68 g of carbohydrate, 150 g of protein, and 9.0 g of fat
Need total energy in kilocalories

Food Type	Mass	Energy Values	Energy
Carbohydrate	68 g	$\times \dfrac{4 \text{ kcal}}{1 \text{ g}} =$	270 kcal
Protein	150 g	$\times \dfrac{4 \text{ kcal}}{1 \text{ g}} =$	600 kcal
Fat	9.0 g	$\times \dfrac{9 \text{ kcal}}{1 \text{ g}} =$	80 kcal
		Total energy content =	950 kcal (energy results are rounded off to tens place)

2.37 **Given** one cup of clam chowder that contains 16 g of carbohydrate, 9 g of protein, and 12 g of fat
Need total energy in kilocalories and kilojoules

Food Type	Mass	Energy Values	Energy
Carbohydrate	16 g	$\times \dfrac{4 \text{ kcal (or 17 kJ)}}{1 \text{ g}} =$	60 kcal (or 270 kJ)
Protein	9 g	$\times \dfrac{4 \text{ kcal (or 17 kJ)}}{1 \text{ g}} =$	40 kcal (or 150 kJ)
Fat	12 g	$\times \dfrac{9 \text{ kcal (or 38 kJ)}}{1 \text{ g}} =$	110 kcal (or 460 kJ)
		Total energy content =	210 kcal (or 880 kJ) (energy results are rounded off to tens place)

2.39 **a.** The change from solid to liquid state is melting.
 b. Coffee is freeze-dried using the process of sublimation.
 c. Liquid water turning to ice is freezing.
 d. Ice crystals form on a package of frozen corn due to deposition.

2.41 **a.** $65 \text{ g ice} \times \dfrac{80. \text{ cal}}{1 \text{ g ice}} = 5200 \text{ cal (2 SFs)}$; heat is absorbed

 b. $17.0 \text{ g ice} \times \dfrac{334 \text{ J}}{1 \text{ g ice}} = 5680 \text{ J (3 SFs)}$; heat is absorbed

 c. $225 \text{ g water} \times \dfrac{80. \text{ cal}}{1 \text{ g water}} \times \dfrac{1 \text{ kcal}}{1000 \text{ cal}} = 18 \text{ kcal (2 SFs)}$; heat is released

 d. $50.0 \text{ g water} \times \dfrac{334 \text{ J}}{1 \text{ g water}} \times \dfrac{1 \text{ kJ}}{1000 \text{ J}} = 16.7 \text{ kJ (3 SFs)}$; heat is released

2.43 **a.** Water vapor in clouds changing to rain is condensation.
 b. Wet clothes drying on a clothesline involves evaporation.
 c. Steam forming as lava flows into the ocean involves boiling.
 d. Water forming on a bathroom mirror after a hot shower involves condensation.

2.45   **a.** $10.0 \text{ g water} \times \dfrac{540 \text{ cal}}{1 \text{ g water}} = 5400 \text{ cal (2 SFs); heat is absorbed}$

   b. $5.00 \text{ g water} \times \dfrac{2260 \text{ J}}{1 \text{ g water}} = 11\,300 \text{ J (3 SFs); heat is absorbed}$

   c. $8.0 \text{ kg steam} \times \dfrac{1000 \text{ g}}{1 \text{ kg}} \times \dfrac{540 \text{ cal}}{1 \text{ g steam}} \times \dfrac{1 \text{ kcal}}{1000 \text{ cal}} = 4300 \text{ kcal (2 SFs); heat is released}$

   d. $175 \text{ g steam} \times \dfrac{2260 \text{ J}}{1 \text{ g steam}} \times \dfrac{1 \text{ kJ}}{1000 \text{ J}} = 396 \text{ kJ (3 SFs); heat is released}$

2.47

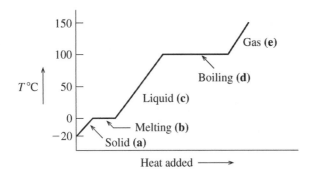

2.49   **a.**   Two calculations are needed:

   (1) ice 0 °C → water 0 °C: $50.0 \text{ g ice} \times \dfrac{334 \text{ J}}{1 \text{ g ice}} = 16\,700 \text{ J}$

   (2) water 0 °C → 65.0 °C: $\Delta T = 65.0 \text{ °C} - 0 \text{ °C} = 65.0 \text{ °C}$

   $50.0 \text{ g water} \times 65.0 \text{ °C} \times \dfrac{4.184 \text{ J}}{\text{g water °C}} = 13\,600 \text{ J}$

   ∴ Total heat needed = 16 700 J + 13 600 J = 30 300 J (3 SFs)

   b.   Two calculations are needed:

   (1) steam 100 °C → water 100 °C: $15.0 \text{ g steam} \times \dfrac{540 \text{ cal}}{1 \text{ g steam}} \times \dfrac{1 \text{ kcal}}{1000 \text{ cal}} = 8.1 \text{ kcal}$

   (2) water 100 °C → 0 °C: $15.0 \text{ g water} \times 100. \text{ °C} \times \dfrac{1.00 \text{ cal}}{\text{g °C}} \times \dfrac{1 \text{ kcal}}{1000 \text{ cal}} = 1.50 \text{ kcal}$

   ∴ Total heat released = 8.1 kcal + 1.50 kcal = 9.6 kcal (2 SFs)

   c.   Three calculations are needed:

   (1) ice 0 °C → water 0 °C: $24.0 \text{ g ice} \times \dfrac{334 \text{ J}}{1 \text{ g ice}} \times \dfrac{1 \text{ kJ}}{1000 \text{ J}} = 8.02 \text{ kJ}$

   (2) water 0 °C → 100 °C: $24.0 \text{ g water} \times 100. \text{ °C} \times \dfrac{4.184 \text{ J}}{\text{g °C}} \times \dfrac{1 \text{ kJ}}{1000 \text{ J}} = 10.0 \text{ kJ}$

   (3) water 100 °C → steam 100 °C: $24.0 \text{ g water} \times \dfrac{2260 \text{ J}}{1 \text{ g water}} \times \dfrac{1 \text{ kJ}}{1000 \text{ J}} = 54.2 \text{ kJ}$

   ∴ Total heat needed = 8.02 kJ + 10.0 kJ + 54.2 kJ = 72.2 kJ (3 SFs)

2.51 **a.** The diagram shows two different types of atoms chemically combined in a specific proportion; it is a compound.

b. The diagram shows two different types of atoms physically mixed, not chemically combined; it is a mixture.

c. The diagram contains only one type of atom; it represents an element.

2.53 A *homogeneous mixture* has a uniform composition; a *heterogeneous mixture* does not have a uniform composition throughout the mixture.

a. Lemon-flavored water is a homogeneous mixture since it has a uniform composition (as long as there are no lemon pieces).

b. Stuffed mushrooms are a heterogeneous mixture since there are mushrooms and chunks of filling.

c. Tortilla soup is a heterogeneous mixture since it contains tortilla pieces and chunks of chicken and vegetables.

2.55 **a.** 61.4 °C

b. 53.80 °C

c. 4.8 °C

2.57 $T_C = \dfrac{(T_F - 32)}{1.8} = \dfrac{(155 - 32)}{1.8} = \dfrac{123}{1.8} = 68.3\ °C$ (1.8 and 32 are exact numbers)

$T_K = T_C + 273 = 68.3 + 273 = 341\ K$ (3 SFs)

2.59 **Given** 10.0 cm³ cubes of gold ($SH_{gold} = 0.129$ J/g °C = 0.0308 cal/g °C), aluminum

($SH_{aluminum} = 0.897$ J/g °C = 0.214 cal/g °C), and silver

($SH_{silver} = 0.235$ J/g °C = 0.0562 cal/g °C); $\Delta T = 25\ °C - 15\ °C = 10.\ °C$

Need energy in joules and calories

Plan cm³ → g   Heat = mass × ΔT × SH

$$\frac{19.3\ g\ gold}{1\ cm^3} \quad \frac{2.70\ g\ aluminum}{1\ cm^3} \quad \frac{10.5\ g\ silver}{1\ cm^3}$$

Set Up for gold: $10.0\ cm^3 \times \dfrac{19.3\ g}{1\ cm^3} \times 10.\ °C \times \dfrac{0.129\ J}{g\ °C} = 250\ J$ (2 SFs)

or $10.0\ cm^3 \times \dfrac{19.3\ g}{1\ cm^3} \times 10.\ °C \times \dfrac{0.0308\ cal}{g\ °C} = 59\ cal$ (2 SFs)

for aluminum: $10.0\ cm^3 \times \dfrac{2.70\ g}{1\ cm^3} \times 10.\ °C \times \dfrac{0.897\ J}{g\ °C} = 240\ J$ (2 SFs)

or $10.0\ cm^3 \times \dfrac{2.70\ g}{1\ cm^3} \times 10.\ °C \times \dfrac{0.214\ cal}{g\ °C} = 58\ cal$ (2 SFs)

for silver: $10.0\ cm^3 \times \dfrac{10.5\ g}{1\ cm^3} \times 10.\ °C \times \dfrac{0.235\ J}{g\ °C} = 250\ J$ (2 SFs)

or $10.0\ cm^3 \times \dfrac{10.5\ g}{1\ cm^3} \times 10.\ °C \times \dfrac{0.0562\ cal}{g\ °C} = 59\ cal$ (2 SFs)

Thus, the heat needed for each of the three samples of metals is almost the same.

2.61 **Given** a meal consisting of: cheeseburger: 31 g of protein, 29 g of fat, 34 g of carbohydrate
french fries: 3 g of protein, 11 g of fat, 26 g of carbohydrate
chocolate shake: 11 g of protein, 9 g of fat, 60 g of carbohydrate

Need total energy in kilocalories

Plan total grams of each food type   then g → kcal

Set Up total protein $= 31$ g $+ 3$ g $+ 11$ g $= 45$ g

total fat $= 29$ g $+ 11$ g $+ 9$ g $= 49$ g

total carbohydrate $= 34$ g $+ 26$ g $+ 60$ g $= 120$ g

Food Type	Mass	Energy Values	Energy
Protein	45 g ×	$\dfrac{4 \text{ kcal}}{1 \text{ g}}$ =	180 kcal
Fat	49 g ×	$\dfrac{9 \text{ kcal}}{1 \text{ g}}$ =	440 kcal
Carbohydrate	120 g ×	$\dfrac{4 \text{ kcal}}{1 \text{ g}}$ =	480 kcal
	Total energy content	=	1100 kcal

(energy results are rounded off to tens place)

2.63 **a.** The heat from the skin is used to evaporate the water (perspiration). Therefore, the skin is cooled.
  b. On a hot day, there are more liquid water molecules in the damp towels that have sufficient energy to become water vapor. Thus, water evaporates from the towels more readily on a hot day.
  c. In a closed plastic bag, some water molecules evaporate, but they cannot escape and will condense back to liquid; the clothes will not dry.

2.65 **a.** The melting point of chloroform is about −60 °C.
  b. The boiling point of chloroform is about 60 °C.
  c. The diagonal line **A** represents the solid state as temperature increases. The horizontal line **B** represents the change from solid to liquid or melting of the substance. The diagonal line **C** represents the liquid state as temperature increases. The horizontal line **D** represents the change from liquid to gas or boiling of the liquid. The diagonal line **E** represents the gas state as temperature increases.
  d. at −80 °C, solid; at −40 °C, liquid; at 25 °C, liquid; at 80 °C, gas

2.67 *Elements* are the simplest type of pure substance, containing only one type of atom. *Compounds* contain two or more elements chemically combined in a specific proportion. In a *mixture*, two or more substances are physically mixed but not chemically combined.
  a. Carbon in pencils is an element since it contains only one type of atom (C).
  b. Carbon dioxide is a compound since it contains two elements (C, O) chemically combined.
  c. Orange juice is composed of several substances mixed together (e.g., water, sugar, citric acid), which makes it a mixture.
  d. Neon gas in lights is an element since it contains only one type of atom (Ne).
  e. A salad dressing made of oil and vinegar is composed of several substances mixed together (e.g., oil, water, acetic acid), which makes it a mixture.

2.69 **a.** A vitamin tablet is a solid.
   b. Helium in a balloon is a gas.
   c. Milk is a liquid.
   d. Air is a mixture of gases.
   e. Charcoal is a solid.

2.71 A *physical property* is a characteristic of the substance such as color, shape, odor, luster, size, melting point, and density. A *chemical property* is a characteristic that indicates the ability of a substance to change into a new substance.
   a. The luster of gold is a physical property.
   b. The melting point of gold is a physical property.
   c. The ability of gold to conduct electricity is a physical property.
   d. The ability of gold to form a new substance with sulfur is a chemical property.

2.73 A *physical change* describes a change in a physical property that retains the identity of the substance. A *chemical change* occurs when the atoms of the initial substances rearrange to form new substances.
   a. Plant growth produces new substances, so it is a chemical change.
   b. A change of state from solid to liquid is a physical change.
   c. Chopping wood into smaller pieces is a physical change.
   d. Burning wood, which forms new substances, is a chemical change.

2.75 **a.** $T_C = \dfrac{(T_F - 32)}{1.8} = \dfrac{(134 - 32)}{1.8} = \dfrac{102}{1.8} = 56.7\ °C$

   b. $T_C = \dfrac{(T_F - 32)}{1.8} = \dfrac{(-69.7 - 32)}{1.8} = \dfrac{-101.7}{1.8} = -56.5\ °C$

2.77 $T_C = \dfrac{(T_F - 32)}{1.8} = \dfrac{(-15 - 32)}{1.8} = \dfrac{-47}{1.8} = -26\ °C$

   $T_K = T_C + 273 = -26 + 273 = 247\ K$

2.79 Sand must have a lower specific heat than water. When both substances absorb the same amount of heat, the final temperature of the sand will be higher than that of water.

2.81 **Given** 1 lb of body fat; 15% (m/m) water in body fat   **Need** kilocalories to "burn off"

   Plan Because each gram of body fat contains 15% water, a person actually loses 85 grams of fat per hundred grams of body fat. (We considered 1 lb of fat as exactly 1 lb.)

   lb of body fat → g of body fat → g of fat → kcal

   $\dfrac{454\text{ g body fat}}{1\text{ lb body fat}} \qquad \dfrac{85\text{ g fat}}{100\text{ g body fat}} \qquad \dfrac{9\text{ kcal}}{1\text{ g fat}}$

   Set Up $1\ \cancel{\text{lb body fat}} \times \dfrac{454\ \cancel{\text{g body fat}}}{1\ \cancel{\text{lb body fat}}} \times \dfrac{85\ \cancel{\text{g fat}}}{100\ \cancel{\text{g body fat}}} \times \dfrac{9\text{ kcal}}{1\ \cancel{\text{g fat}}} = 3500\text{ kcal (2 SFs)}$

2.83 **Given** mass = 725 g of water; $SH_{water} = 1.00$ cal/g °C; $\Delta T = 37\ °C - 65\ °C = -28\ °C$

   Need heat in kilocalories

   Plan Heat = mass $\times \Delta T \times SH$ then cal → kcal $\dfrac{1\text{ kcal}}{1000\text{ cal}}$

Set Up $\text{Heat} = 725 \cancel{g} \times (-28\,\cancel{°C}) \times \dfrac{1.00\,\cancel{cal}}{\cancel{g}\,\cancel{°C}} \times \dfrac{1\,\text{kcal}}{1000\,\cancel{cal}} = -20.\text{ kcal (2 SFs)}$

∴ 20. kcal of heat are lost from the water bottle and could be transferred to sore muscles.

2.85

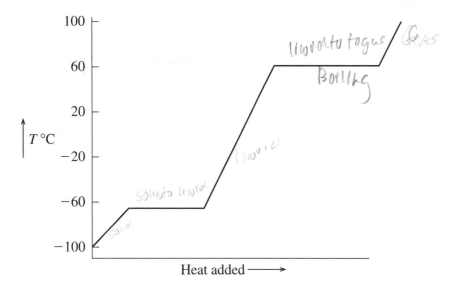

   a. Chloroform is a solid at −75 °C.
   b. At −64 °C, chloroform melts (solid → liquid).
   c. Chloroform is a liquid at −18 °C.
   d. At 80 °C, chloroform is a gas.
   e. Both solid and liquid chloroform will be present at the melting temperature of −64 °C.

2.87   Convert 140. °F to °C:

$$T_C = \frac{(T_F - 32)}{1.8} = \frac{(140. - 32)}{1.8} = \frac{108}{1.8} = 60.0\ °C$$

∴ the two liquids are at the same temperature, 60.0 °C (140. °F).

2.89   Three calculations are needed:

(1) steam 100 °C → water 100 °C: $75\ \cancel{\text{g steam}} \times \dfrac{540\,\cancel{cal}}{1\ \cancel{\text{g steam}}} \times \dfrac{1\,\text{kcal}}{1000\,\cancel{cal}} = 41\ \text{kcal}$

(2) water 100 °C → 0 °C: $\Delta T = 100\ °C - 0\ °C = 100.\ °C$

$75\ \cancel{\text{g water}} \times 100.\,\cancel{°C} \times \dfrac{1.00\,\cancel{cal}}{\text{g water}\ \cancel{°C}} \times \dfrac{1\,\text{kcal}}{1000\,\cancel{cal}} = 7.5\ \text{kcal}$

(3) water 0 °C → ice 0 °C: $75\ \cancel{\text{g water}} \times \dfrac{80.\,\cancel{cal}}{1\ \cancel{\text{g water}}} \times \dfrac{1\,\text{kcal}}{1000\,\cancel{cal}} = 6.0\ \text{kcal}$

∴ Total heat released = 41 kcal + 7.5 kcal + 6.0 kcal = 55 kcal (2 SFs)

2.91   **a.** **Given** 2.4×10^7 J/1.0 lb of oil; mass $=150$ kg of water; $SH_{\text{water}}=4.184$ J/g °C;

$$\Delta T = 100\ ^\circ\text{C} - 22\ ^\circ\text{C} = 78\ ^\circ\text{C}$$

Need kilograms of oil needed

Plan Heat $=$ mass $\times \Delta T \times SH$ then J $\rightarrow$ lb of oil $\rightarrow$ kg of oil $\dfrac{1\text{ kg}}{2.20\text{ lb}}$

Set Up Heat $=150$ kg water $\times \dfrac{1000\ \cancel{g}}{1\ \cancel{kg}} \times 78\ \cancel{^\circ C} \times \dfrac{4.184\text{ J}}{\cancel{g}\ \cancel{^\circ C}} = 4.9\times10^7$ J (2 SFs)

$\therefore$ mass of oil $= 4.9\times10^7\ \cancel{J} \times \dfrac{1.0\ \cancel{\text{lb oil}}}{2.4\times10^7\ \cancel{J}} \times \dfrac{1.0\text{ kg oil}}{2.20\ \cancel{\text{lb oil}}} = 0.93$ kg of oil (2 SFs)

b. **Given** 2.4×10^7 J/1.0 lb of oil; mass $=150$ kg of water; water 100 °C $\rightarrow$ steam 100 °C

Need kilograms of oil needed

Plan heat for water 100 °C $\rightarrow$ steam 100 °C;   then J $\rightarrow$ lb of oil $\rightarrow$ kg of oil $\dfrac{1\text{ kg}}{2.20\text{ lb}}$

Set Up Heat $=150$ kg water $\times \dfrac{1000\ \cancel{g}}{1\ \cancel{kg}} \times \dfrac{2260\text{ J}}{1\text{ g water}} = 3.4\times10^8$ J (2 SFs)

$\therefore$ mass of oil $= 3.4\times10^8\ \cancel{J} \times \dfrac{1.0\ \cancel{\text{lb oil}}}{2.4\times10^7\ \cancel{J}} \times \dfrac{1.0\text{ kg oil}}{2.20\ \cancel{\text{lb oil}}} = 6.4$ kg of oil (2 SFs)

2.93   Two calculations are required:

(1) ice 0 °C $\rightarrow$ water 0 °C: $275\ \cancel{\text{g ice}} \times \dfrac{334\ \cancel{J}}{1\ \cancel{\text{g ice}}} \times \dfrac{1\text{ kJ}}{1000\ \cancel{J}} = 91.9$ kJ

(2) water 0 °C $\rightarrow$ 24.0 °C: $\Delta T = 24.0\ ^\circ\text{C} - 0\ ^\circ\text{C} = 24.0\ ^\circ\text{C}$;

$275\ \cancel{\text{g water}} \times 24.0\ \cancel{^\circ C} \times \dfrac{4.184\ \cancel{J}}{\cancel{g}\ \cancel{^\circ C}} \times \dfrac{1\text{ kJ}}{1000\ \cancel{J}} = 27.6$ kJ

$\therefore$ Total heat absorbed $= 91.9$ kJ $+ 27.6$ kJ $= 119.5$ kJ

Answers to Combining Ideas from Chapters 1 and 2

CI.1   **a.**   There are 4 significant figures in 20.17 lb.

     b. $20.17 \text{ lb} \times \dfrac{1 \text{ kg}}{2.20 \text{ lb}} = 9.17 \text{ kg (3 SFs)}$

     c. $9.17 \text{ kg} \times \dfrac{1000 \text{ g}}{1 \text{ kg}} \times \dfrac{1 \text{ cm}^3}{19.3 \text{ g}} = 475 \text{ cm}^3 \text{ (3 SFs)}$

     d. $T_F = 1.8(T_C) + 32 = 1.8(1064) + 32 = 1947 \text{ °F (4 SFs)}$

         $T_K = T_C + 273 = 1064 + 273 = 1337 \text{ K (4 SFs)}$

     e. $\Delta T = T_{final} - T_{initial} = 1064 \text{ °C} - 500. \text{ °C} = 564 \text{ °C} = \text{temperature change}$

$$9.17 \text{ kg} \times \dfrac{1000 \text{ g}}{1 \text{ kg}} \times \dfrac{0.0308 \text{ cal}}{\text{g °C}} \times 564 \text{ °C} \times \dfrac{1 \text{ kcal}}{1000 \text{ cal}} = 159 \text{ kcal to raise the temperature}$$

$$9.17 \text{ kg} \times \dfrac{1000 \text{ g}}{1 \text{ kg}} \times \dfrac{63.6 \text{ J}}{\text{g}} \times \dfrac{1 \text{ cal}}{4.184 \text{ J}} \times \dfrac{1 \text{ kcal}}{1000 \text{ cal}} = 139 \text{ kcal to melt}$$

Total energy $= 159 \text{ kcal} + 139 \text{ kcal} = 298 \text{ kcal (3 SFs)}$

     f. $9.17 \text{ kg} \times \dfrac{1000 \text{ g}}{1 \text{ kg}} \times \dfrac{\$35.10}{1 \text{ g}} = \$322\,000 \text{ (3 SFs)}$

CI.3   **a.**   The shape of **A** changes to the shape of the new container, while the shape of **B** remains the same. The volumes of both **A** and **B** remain the same.

     b.   **A** is liquid water represented by diagram 2. In liquid water, the water particles are in a random arrangement, but close together. **B** is solid water represented by diagram 1. In solid water, the water particles are fixed in a definite arrangement.

     c.   solid; liquid; gas

     d.   melting; melting point; physical

     e.   boiling; boiling point; physical

     f.   Two calculations are needed:

(1) water 45 °C $\rightarrow$ 0 °C: $\Delta T = 45 \text{ °C} - 0 \text{ °C} = 45 \text{ °C}$

$$19 \text{ g water} \times 45 \text{ °C} \times \dfrac{4.184 \text{ J}}{\text{g water °C}} \times \dfrac{1 \text{ kJ}}{1000 \text{ J}} = 3.6 \text{ kJ}$$

(2) water 0 °C $\rightarrow$ ice 0 °C: $19 \text{ g water} \times \dfrac{334 \text{ J}}{1 \text{ g water}} \times \dfrac{1 \text{ kJ}}{1000 \text{ J}} = 6.3 \text{ kJ}$

$\therefore$ Total heat released $= 3.6 \text{ kJ} + 6.3 \text{ kJ} = 9.9 \text{ kJ (2 SFs)}$

CI.5   **a.** $0.25 \text{ lb} \times \dfrac{454 \text{ g}}{1 \text{ lb}} \times \dfrac{1 \text{ cm}^3}{7.86 \text{ g}} = 14 \text{ cm}^3$ (2 SFs)

   b. $30 \text{ nails} \times \dfrac{0.25 \text{ lb}}{75 \text{ nails}} \times \dfrac{454 \text{ g}}{1 \text{ lb}} \times \dfrac{1 \text{ cm}^3}{7.86 \text{ g}} \times \dfrac{1 \text{ mL}}{1 \text{ cm}^3} = 5.8 \text{ mL}$

   17.6 mL water + 5.8 mL = 23.4 mL new water level (3 SFs)

   c. $\Delta T = T_{\text{final}} - T_{\text{initial}} = 125\ ^\circ\text{C} - 16\ ^\circ\text{C} = 109\ ^\circ\text{C} = \text{temperature change}$

   $\text{Heat} = \text{mass} \times \Delta T \times SH = 0.25 \text{ lb} \times \dfrac{454 \text{ g}}{1 \text{ lb}} \times 109\ ^\circ\text{C} \times \dfrac{0.452 \text{ J}}{\text{g } ^\circ\text{C}} = 5600 \text{ J or } 5.6 \times 10^3 \text{ J (3 SFs)}$

   d.   Two calculations are needed:

   (1) iron 25 °C → 1535 °C: $\Delta T = 1535\ ^\circ\text{C} - 25\ ^\circ\text{C} = 1510\ ^\circ\text{C}$

   $1 \text{ nail} \times \dfrac{0.25 \text{ lb Fe}}{75 \text{ nails}} \times \dfrac{454 \text{ g Fe}}{1 \text{ lb Fe}} \times \dfrac{0.452 \text{ J}}{\text{g } ^\circ\text{C}} \times 1510.\ ^\circ\text{C} = 1.0 \times 10^3 \text{ J}$

   (2) solid iron 1535 °C → liquid iron 1535 °C:

   $1 \text{ nail} \times \dfrac{0.25 \text{ lb Fe}}{75 \text{ nails}} \times \dfrac{454 \text{ g Fe}}{1 \text{ lb Fe}} \times \dfrac{272 \text{ J}}{1 \text{ g Fe}} = 410 \text{ J}$

   $\therefore$ Total heat required $= 1.0 \times 10^3 \text{ J} + 410 \text{ J} = 1.4 \times 10^3 \text{ J (2 SFs)}$

3.1   **a.** Cu is the symbol for copper.
     b. Pt is the symbol for platinum.
     c. Ca is the symbol for calcium.
     d. Mn is the symbol for manganese.
     e. Fe is the symbol for iron.
     f. Ba is the symbol for barium.
     g. Pb is the symbol for lead.
     h. Sr is the symbol for strontium.

3.3   **a.** Carbon is the element with the symbol C.
     b. Chlorine is the element with the symbol Cl.
     c. Iodine is the element with the symbol I.
     d. Mercury is the element with the symbol Hg.
     e. Silver is the element with the symbol Ag.
     f. Argon is the element with the symbol Ar.
     g. Boron is the element with the symbol B.
     h. Nickel is the element with the symbol Ni.

3.5   **a.** Sodium (Na) and chlorine (Cl) are in NaCl.
     b. Calcium (Ca), sulfur (S), and oxygen (O) are in $CaSO_4$.
     c. Carbon (C), hydrogen (H), chlorine (Cl), nitrogen (N), and oxygen (O) are in $C_{15}H_{22}ClNO_2$.
     d. Calcium (Ca), carbon (C), and oxygen (O) are in $CaCO_3$.

3.7   **a.** C, N, and O are in Period 2.
     b. He is the element at the top of Group 8A (18).
     c. The alkali metals are the elements in Group 1A (1).
     d. Period 2 is the horizontal row of elements that ends with neon (Ne).

3.9   **a.** Ca is an alkaline earth metal.
     b. Fe is a transition element.
     c. Xe is a noble gas.
     d. K is an alkali metal.
     e. Cl is a halogen.

3.11 **a.** C is Group 4A (14), Period 2.
     b. He is the noble gas in Period 1.
     c. Na is the alkali metal in Period 3.
     d. Ca is Group 2A (2), Period 4.
     e. Al is Group 3A (13), Period 3.

3.13 On the periodic table, *metals* are located to the left of the heavy zigzag line, *nonmetals* are elements to the right, and *metalloids* (B, Si, Ge, As, Sb, Te, Po, and At) are located along the line.
     a. Ca is a metal.
     b. S is a nonmetal.
     c. Metals are shiny.
     d. An element that is a gas at room temperature is a nonmetal.

     e. Group 8A (18) elements are nonmetals.
     f. Br is a nonmetal.
     g. Te is a metalloid.
     h. Ag is a metal.

3.15   **a.** The electron has the smallest mass.
     b. The proton has a 1+ charge.
     c. The electron is found outside the nucleus.
     d. The neutron is electrically neutral.

3.17 Rutherford determined that positively charged particles called protons are located in a very small, dense region of the atom called the nucleus.

3.19   **a.** True
     b. True
     c. True
     d. False; since a neutron has no charge, it is not attracted to a proton (a proton is attracted to an electron).

3.21 In the process of brushing hair, strands of hair become charged with like charges that repel each other.

3.23   **a.** The atomic number is the same as the number of protons in an atom.
     b. Both are needed since the number of neutrons is the (mass number) − (atomic number).
     c. The mass number is the number of particles (protons) + (neutrons) in the nucleus.
     d. The atomic number is the same as the number of electrons in a neutral atom.

3.25 The atomic number defines the element and is found above the symbol of the element in the periodic table.
     a. Lithium, Li, has an atomic number of 3.
     b. Fluorine, F, has an atomic number of 9.
     c. Calcium, Ca, has an atomic number of 20.
     d. Zinc, Zn, has an atomic number of 30.
     e. Neon, Ne, has an atomic number of 10.
     f. Silicon, Si, has an atomic number of 14.
     g. Iodine, I, has an atomic number of 53.
     h. Oxygen, O, has an atomic number of 8.

3.27 The atomic number gives the number of protons in the nucleus of an atom. Since atoms are neutral, the atomic number also gives the number of electrons in the neutral atom.
     a. There are 18 protons and 18 electrons in a neutral argon atom.
     b. There are 30 protons and 30 electrons in a neutral zinc atom.
     c. There are 53 protons and 53 electrons in a neutral iodine atom.
     d. There are 19 protons and 19 electrons in a neutral potassium atom.

3.29 The atomic number is the same as the number of protons in an atom and the number of electrons in a neutral atom; the atomic number defines the element. The number of neutrons is the (mass number) − (atomic number).

Name of Element	Symbol	Atomic Number	Mass Number	Number of Protons	Number of Neutrons	Number of Electrons
Aluminum	Al	13	27	13	$27 - 13 = 14$	13
Magnesium	Mg	12	$12 + 12 = 24$	12	12	12
Potassium	K	19	$19 + 20 = 39$	19	20	19
Sulfur	S	16	$16 + 15 = 31$	16	15	16
Iron	Fe	26	56	26	$56 - 26 = 30$	26

3.31   **a.** Since the atomic number of aluminum is 13, every Al atom has 13 protons. An atom of aluminum (mass number 27) has 14 neutrons ($27 - 13 = 14 \, n$). Neutral atoms have the same number of protons and electrons. Therefore, 13 protons, 14 neutrons, 13 electrons.

   b. Since the atomic number of chromium is 24, every Cr atom has 24 protons. An atom of chromium (mass number 52) has 28 neutrons ($52 - 24 = 28 \, n$). Neutral atoms have the same number of protons and electrons. Therefore, 24 protons, 28 neutrons, 24 electrons.

   c. Since the atomic number of sulfur is 16, every S atom has 16 protons. An atom of sulfur (mass number 34) has 18 neutrons ($34 - 16 = 18 \, n$). Neutral atoms have the same number of protons and electrons. Therefore, 16 protons, 18 neutrons, 16 electrons.

   d. Since the atomic number of bromine is 35, every Br atom has 35 protons. An atom of bromine (mass number 81) has 46 neutrons ($81 - 35 = 46 \, n$). Neutral atoms have the same number of protons and electrons. Therefore, 35 protons, 46 neutrons, 35 electrons.

3.33   **a.** Since the number of protons is 15, the atomic number is 15 and the element symbol is P. The mass number is the sum of the number of protons and the number of neutrons, $15 + 16 = 31$. The atomic symbol for this isotope is $^{31}_{15}\text{P}$.

   b. Since the number of protons is 35, the atomic number is 35 and the element symbol is Br. The mass number is the sum of the number of protons and the number of neutrons, $35 + 45 = 80$. The atomic symbol for this isotope is $^{80}_{35}\text{Br}$.

   c. Since the number of electrons is 50, there must be 50 protons in a neutral atom. Since the number of protons is 50, the atomic number is 50, and the element symbol is Sn. The mass number is the sum of the number of protons and the number of neutrons, $50 + 72 = 122$. The atomic symbol for this isotope is $^{122}_{50}\text{Sn}$.

   d. Since the element is chlorine, the element symbol is Cl, the atomic number is 17, and the number of protons is 17. The mass number is the sum of the number of protons and the number of neutrons, $17 + 18 = 35$. The atomic symbol for this isotope is $^{35}_{17}\text{Cl}$.

   e. Since the element is mercury, the element symbol is Hg, the atomic number is 80, and the number of protons is 80. The mass number is the sum of the number of protons and the number of neutrons, $80 + 122 = 202$. The atomic symbol for this isotope is $^{202}_{80}\text{Hg}$.

3.35   **a.** Since the element is argon, the element symbol is Ar, the atomic number is 18, and the number of protons is 18. The atomic symbols for the isotopes with mass numbers of 36, 38, and 40 are $^{36}_{18}\text{Ar}$, $^{38}_{18}\text{Ar}$, and $^{40}_{18}\text{Ar}$, respectively.

   b. They all have the same atomic number (the same number of protons and electrons).

   c. They have different numbers of neutrons, which gives them different mass numbers.

   d. The atomic mass of argon listed on the periodic table is the weighted average atomic mass of all the naturally occurring isotopes of argon.

   e. The isotope Ar-40 ($^{40}_{18}\text{Ar}$) is the most prevalent in a sample of argon because its mass is closest to the average atomic mass of argon listed on the periodic table (39.95 amu).

3.37 $^{69}_{31}$Ga $68.93 \times \dfrac{60.11}{100} = 41.43$ amu

        $^{71}_{31}$Ga $70.92 \times \dfrac{39.89}{100} = 28.29$ amu

        Atomic mass of Ga = 69.72 amu (4 SFs)

3.39   Electrons can move to higher energy levels when they <u>absorb</u> energy.

3.41   **a.**   Green light has greater energy than yellow light.
        b.   Blue light has greater energy than microwaves.

3.43   **a.**   Carbon with atomic number 6 has 6 electrons, which are arranged with 2 electrons in energy level 1 and 4 electrons in energy level 2:   2,4
        b.   Argon with atomic number 18 has 18 electrons, which are arranged with 2 electrons in energy level 1, 8 electrons in energy level 2, and 8 electrons in energy level 3:   2,8,8
        c.   Potassium with atomic number 19 has 19 electrons, which are arranged with 2 electrons in energy level 1, 8 electrons in energy level 2, 8 electrons in energy level 3, and 1 electron in energy level 4:   2,8,8,1
        d.   Silicon with atomic number 14 has 14 electrons, which are arranged with 2 electrons in energy level 1, 8 electrons in energy level 2, and 4 electrons in energy level 3:   2,8,4
        e.   Helium with atomic number 2, has 2 electrons, which are arranged with 2 electrons in energy level 1:   2
        f.   Nitrogen with atomic number 7 has 7 electrons, which are arranged with 2 electrons in energy level 1 and 5 electrons in energy level 2:   2,5

3.45   **a.**   2,1 is the electron arrangement of lithium (Li).
        b.   2,8,2 is the electron arrangement of magnesium (Mg).
        c.   1 is the electron arrangement of hydrogen (H).
        d.   2,8,7 is the electron arrangement of chlorine (Cl).
        e.   2,6 is the electron arrangement of oxygen (O).

3.47   **a.**   Magnesium, in Group 2A (2), has 2 valence electrons.
        b.   Chlorine, in Group 7A (17), has 7 valence electrons.
        c.   Oxygen, in Group 6A (16), has 6 valence electrons.
        d.   Nitrogen, in Group 5A (15), has 5 valence electrons.
        e.   Barium, in Group 2A (2), has 2 valence electrons.
        f.   Bromine, in Group 7A (17), has 7 valence electrons.

3.49   **a.**   Sulfur is in Group 6A (16);

           $\cdot\overset{\displaystyle\cdot\cdot}{\underset{\displaystyle\cdot}{S}}\!:$

        b.   Nitrogen is in Group 5A (15);

           $\cdot\overset{\displaystyle\cdot\cdot}{\underset{\displaystyle\cdot}{N}}\cdot$

        c.   Calcium is in Group 2A (2);

           $\overset{\displaystyle\cdot}{Ca}\cdot$

        d.   Sodium is in Group 1A (1);

           $Na\cdot$

        e.   Gallium is in Group 3A (13);

           $\cdot\overset{\displaystyle\cdot}{Ga}\cdot$

3.51 **a.** The atomic radius of representative elements decreases from Group 1A to 8A: Mg, Al, Si.
         b. The atomic radius of representative elements increases going down a group: I, Br, Cl.
         c. The atomic radius of representative elements decreases from Group 1A to 8A: Sr, Sb, I.
         d. The atomic radius of representative elements decreases from Group 1A to 8A: Na, Si, P.

3.53 The atomic radius of representative elements decreases going across a period from Group 1A to 8A and increases going down a group.
         a. In Period 3, Na, which is on the left, is larger than Cl.
         b. In Group 1A (1), Rb, which is farther down the group, is larger than Na.
         c. In Period 3, Na, which is on the left, is larger than Mg.
         d. In Period 5, Rb, which is on the left, is larger than I.

3.55 **a.** Br, Cl, F; ionization energy decreases going down a group.
         b. Na, Al, Cl; going across a period from left to right, ionization energy increases.
         c. Cs, K, Na; ionization energy decreases going down a group.
         d. Ca, As, Br; going across a period from left to right, ionization energy increases.

3.57 **a.** In Br, the valence electrons are closer to the nucleus, so Br has a higher ionization energy than I.
         b. In Mg, the valence electrons are closer to the nucleus, so Mg has a higher ionization energy than Sr.
         c. Attraction for the valence electrons increases going from left to right across a period, giving P a higher ionization energy than Si.
         d. The noble gases have the highest ionization energy in each period, which gives Xe a higher ionization energy than I.

3.59 Na has a <u>larger</u> atomic size and is <u>more metallic</u> than P.

3.61 Since metallic character decreases from left to right across a period, Ca, in Group 2A (2), will be the most metallic, followed by Ga, then Ge, and finally Br, in Group 7A (17), will have the least metallic character of the elements listed.

3.63 Sr has a <u>lower</u> ionization energy and <u>more</u> metallic character than Sb.

3.65 Going down Group 6A (16),
         a. the ionization energy <u>decreases</u>
         b. the atomic size <u>increases</u>
         c. the metallic character <u>increases</u>
         d. the number of valence electrons <u>remains the same</u>

3.67 In Period 2, the nonmetals compared to the metals have larger (greater)
         b. ionization energies
         c. number of protons
         e. number of valence electrons
         (options **a**, **d** are false)

3.69 Statements **b** and **c** are true. According to Dalton's atomic theory, atoms of an element are different than atoms of other elements, and atoms do not appear and disappear in a chemical reaction.

3.71 **a.** The atomic mass is the weighted average of the masses of all of the naturally occurring isotopes of the element. Isotope masses are based on the masses of all subatomic particles in the atom: protons (1), neutrons (2), and electrons (3), although almost all of that mass comes from the protons (1) and neutrons (2).
         b. The number of protons (1) is the atomic number.
         c. The protons (1) are positively charged.
         d. The electrons (3) are negatively charged.
         e. The number of neutrons (2) is the (mass number) – (atomic number).

3.73 **a.** $^{16}_{8}X$, $^{17}_{8}X$, and $^{18}_{8}X$ all have an atomic number of 8, so all have 8 protons.

b. $^{16}_{8}X$, $^{17}_{8}X$, and $^{18}_{8}X$ all have an atomic number of 8, so all are isotopes of oxygen.

c. $^{16}_{8}X$ and $^{16}_{9}X$ have mass numbers of 16, whereas $^{18}_{8}X$ and $^{18}_{10}X$ have mass numbers of 18.

d. $^{16}_{8}X$ $(16 - 8 = 8\ n)$ and $^{18}_{10}X$ $(18 - 10 = 8\ n)$ both have 8 neutrons.

3.75 **a.** $^{37}_{17}Cl$ and $^{38}_{18}Ar$ both have 20 neutrons $(37 - 17 = 20\ n;\ 38 - 18 = 20\ n)$.

b. Since both $^{36}_{14}Si$ and $^{35}_{14}Si$ have an atomic number of 14, they will both have 14 protons in the nucleus. Since the number of protons is 14, there must be 14 electrons in the neutral atom for both. They have different numbers of neutrons $(36 - 14 = 22\ n;\ 35 - 14 = 21\ n)$.

c. $^{40}_{18}Ar$ and $^{39}_{17}Cl$ both have 22 neutrons $(40 - 18 = 22\ n;\ 39 - 17 = 22\ n)$.

3.77 **a.** The diagram shows 4 protons and 5 neutrons in the nucleus of the element. Since the number of protons is 4, the atomic number is 4, and the element symbol is Be. The mass number is the sum of the number of protons and the number of neutrons, $4 + 5 = 9$. The atomic symbol is $^{9}_{4}Be$.

b. The diagram shows 5 protons and 6 neutrons in the nucleus of the element. Since the number of protons is 5, the atomic number is 5, and the element symbol is B. The mass number is the sum of the number of protons and the number of neutrons, $5 + 6 = 11$. The atomic symbol is $^{11}_{5}B$.

c. The diagram shows 6 protons and 7 neutrons in the nucleus of the element. Since the number of protons is 6, the atomic number is 6, and the element symbol is C. The mass number is the sum of the number of protons and the number of neutrons, $6 + 7 = 13$. The atomic symbol is $^{13}_{6}C$.

d. The diagram shows 5 protons and 5 neutrons in the nucleus of the element. Since the number of protons is 5, the atomic number is 5, and the element symbol is B. The mass number is the sum of the number of protons and the number of neutrons, $5 + 5 = 10$. The atomic symbol is $^{10}_{5}B$.

e. The diagram shows 6 protons and 6 neutrons in the nucleus of the element. Since the number of protons is 6, the atomic number is 6, and the element symbol is C. The mass number is the sum of the number of protons and the number of neutrons, $6 + 6 = 12$. The atomic symbol is $^{12}_{6}C$.

Both **b** ($^{11}_{5}B$) and **d** ($^{10}_{5}B$) are isotopes of boron; **c** ($^{11}_{6}C$) and **e** ($^{12}_{6}C$) are isotopes of carbon.

3.79 Atomic radius increases going down a group. Li is **D** because it would be smallest. Na is **A**, K is **C**, and Rb is **B**.

3.81 **a.** Na has the largest atomic size.
b. Cl is a halogen.
c. Si has the electron arrangement 2,8,4.
d. Ar, a noble gas, has the highest ionization energy.
e. S is in Group 6A (16).
f. Na, in Group 1A (1), has the most metallic character.
g. Mg, in Group 2A (2), has two valence electrons.

3.83 **a.** Br is in Group 7A (17), Period 4.
b. Ar is in Group 8A (18), Period 3.
c. K is in Group 1A (1), Period 4.
d. Ra is in Group 2A (2), Period 7.

3.85
    a. False. A proton is a positively charged particle.
    b. False. The neutron has about the same mass as a proton.
    c. True
    d. False. The nucleus is the tiny, dense central core of an atom.
    e. True

3.87
    a. The atomic number gives the number of <u>protons</u> in the nucleus.
    b. In an atom, the number of electrons is equal to the number of <u>protons</u>.
    c. Sodium and potassium are examples of elements called <u>alkali metals</u>.

3.89 The atomic number defines the element and is found above the symbol of the element in the periodic table.
    a. Nickel, Ni, has an atomic number of 28.
    b. Barium, Ba, has an atomic number of 56.
    c. Radium, Ra, has an atomic number of 88.
    d. Arsenic, As, has an atomic number of 33.
    e. Tin, Sn, has an atomic number of 50.
    f. Cesium, Cs, has an atomic number of 55.
    g. Gold, Au, has an atomic number of 79.
    h. Mercury, Hg, has an atomic number of 80.

3.91
    a. Since the atomic number of silver is 47, every Ag atom has 47 protons. An atom of silver (mass number 107) has 60 neutrons ($107 - 47 = 60\ n$). Neutral atoms have the same number of protons and electrons. Therefore, 47 protons, 60 neutrons, 47 electrons.
    b. Since the atomic number of technetium is 43, every Tc atom has 43 protons. An atom of technetium (mass number 98) has 55 neutrons ($98 - 43 = 55\ n$). Neutral atoms have the same number of protons and electrons. Therefore, 43 protons, 55 neutrons, 43 electrons.
    c. Since the atomic number of lead is 82, every Pb atom has 82 protons. An atom of lead (mass number 208) has 126 neutrons ($208 - 82 = 126\ n$). Neutral atoms have the same number of protons and electrons. Therefore, 82 protons, 126 neutrons, 82 electrons.
    d. Since the atomic number of radon is 86, every Rn atom has 86 protons. An atom of radon (mass number 222) has 136 neutrons ($222 - 86 = 136\ n$). Neutral atoms have the same number of protons and electrons. Therefore, 86 protons, 136 neutrons, 86 electrons.
    e. Since the atomic number of xenon is 54, every Xe atom has 54 protons. An atom of xenon (mass number 136) has 82 neutrons ($136 - 54 = 82\ n$). Neutral atoms have the same number of protons and electrons. Therefore, 54 protons, 82 neutrons, 54 electrons.

3.93

Name	Atomic Symbol	Number of Protons	Number of Neutrons	Number of Electrons
Sulfur	$^{34}_{16}\text{S}$	16	$34 - 16 = 18$	16
Nickel	$^{28+34}_{28}\text{Ni}$ or $^{62}_{28}\text{Ni}$	28	34	28
Magnesium	$^{12+14}_{12}\text{Mg}$ or $^{26}_{12}\text{Mg}$	12	14	12
Radon	$^{220}_{86}\text{Rn}$	86	$220 - 86 = 134$	86

3.95
    a. Since the number of protons is 4, the atomic number is 4, and the element symbol is Be. The mass number is the sum of the number of protons and the number of neutrons, $4 + 5 = 9$. The atomic symbol for this isotope is $^{9}_{4}\text{Be}$.
    b. Since the number of protons is 12, the atomic number is 12, and the element symbol is Mg. The mass number is the sum of the number of protons and the number of neutrons, $12 + 14 = 26$. The atomic symbol for this isotope is $^{26}_{12}\text{Mg}$.

    c. Since the element is calcium, the element symbol is Ca, and the atomic number is 20. The mass number is given as 46. The atomic symbol for this isotope is $^{46}_{20}$Ca.

    d. Since the number of electrons is 30, there must be 30 protons in a neutral atom. Since the number of protons is 30, the atomic number is 30, and the element symbol is Zn. The mass number is the sum of the number of protons and the number of neutrons, $30 + 40 = 70$. The atomic symbol for this isotope is $^{70}_{30}$Zn.

    e. Since the element is copper, the element symbol is Cu, the atomic number is 29, and the number of protons is 29. The mass number is the sum of the number of protons and the number of neutrons, $29 + 34 = 63$. The atomic symbol for this isotope is $^{63}_{29}$Cu.

3.97   **a.** Since the element is lead (Pb), the atomic number is 82, and the number of protons is 82. In a neutral atom, the number of electrons is equal to the number of protons, so there will be 82 electrons. The number of neutrons is the (mass number) − (atomic number) = $208 - 82 = 126 \, n$. Therefore, 82 protons, 126 neutrons, 82 electrons.

    b. Since the element is lead (Pb), the atomic number is 82, and the number of protons is 82. The mass number is the sum of the number of protons and the number of neutrons, $82 + 132 = 214$. The atomic symbol for this isotope is $^{214}_{82}$Pb.

    c. Since the mass number is 214 (as in part **b**) and the number of neutrons is 131, the number of protons is the (mass number) − (number of neutrons) = $214 - 131 = 83 \, p$. Since there are 83 protons, the atomic number is 83, and the element symbol is Bi (bismuth). The atomic symbol for this isotope is $^{214}_{83}$Bi.

3.99   **a.** Oxygen is in Group 6A (16) and has an electron arrangement of 2,6.

    b. Sodium is in Group 1A (1) and has an electron arrangement of 2,8,1.

    c. Neon is in Group 8A (18) and has an electron arrangement of 2,8.

    d. Boron is in Group 3A (13) and has an electron arrangement of 2,3.

3.101 Ca has a greater number of protons than K; this greater nuclear charge makes it more difficult to remove an electron from Ca. The least tightly bound (valence) electron in Ca is farther from the nucleus than in Mg and less energy is needed to remove it.

3.103 **a.** Be is an alkaline earth metal.

    b. Li has the largest atomic radius.

    c. F, in Group 7A (17), has the highest ionization energy.

    d. N is in Group 5A (15).

    e. Li has the most metallic character.

3.105 **a.** Since the atomic number is 17, the element is chlorine (Cl). Every Cl atom has 17 protons. An atom of chlorine (mass number 37) has 20 neutrons ($37 - 17 = 20 \, n$).

    b. Since the atomic number is 27, the element is cobalt (Co). Every Co atom has 27 protons. An atom of cobalt (mass number 56) has 29 neutrons ($56 - 27 = 29 \, n$).

    c. Since the atomic number is 50, the element is tin (Sn). Every Sn atom has 50 protons. An atom of tin (mass number 116) has 66 neutrons ($116 - 50 = 66 \, n$).

    d. Since the atomic number is 50, the element is tin (Sn). Every Sn atom has 50 protons. An atom of tin (mass number 124) has 74 neutrons ($124 - 50 = 74 \, n$).

    e. Since the atomic number is 48, the element is cadmium (Cd). Every Cd atom has 48 protons. An atom of cadmium (mass number 116) has 68 neutrons ($116 - 48 = 68 \, n$).

    Both **c** ($^{116}_{50}$Sn) and **d** ($^{124}_{50}$Sn) are isotopes of tin.

3.107 **a.** Atomic radius increases going down a group, which gives O the smallest atomic size in Group 6A.

b. Atomic radius decreases going from left to right across a period, which gives Ar the smallest atomic size in Period 3.

c. Ionization energy decreases going down a group, which gives N the highest ionization energy in Group 5A (15).

d. Ionization energy increases going from left to right across a period, which gives Na the lowest ionization energy in Period 3.

e. Metallic character increases going down a group, which gives Ra the most metallic character in Group 2A (2).

3.109 **Given** volume $= 2.00$ cm^3, lead atoms (3.4×10^{-22} g each)   **Need** number of Pb atoms

Plan cm$^3 \to$ g $\to$ number of Pa atoms $\quad \dfrac{11.3 \text{ g}}{1 \text{ cm}^3} \quad \dfrac{1 \text{ atom Pb}}{3.4 \times 10^{-22} \text{ g}}$

Set Up $2.00 \ \cancel{\text{cm}^3} \times \dfrac{11.3 \ \cancel{\text{g}}}{1 \ \cancel{\text{cm}^3}} \times \dfrac{1 \text{ atom Pb}}{3.4 \times 10^{-22} \ \cancel{\text{g}}} = 6.6 \times 10^{22}$ atoms of Pb (2 SFs)

3.111

$^{28}_{14}\text{Si}$ $27.977 \times \dfrac{92.23}{100} = 25.80$ amu

$^{29}_{14}\text{Si}$ $28.976 \times \dfrac{4.68}{100} = 1.36$ amu

$^{30}_{14}\text{Si}$ $29.974 \times \dfrac{3.09}{100} = \underline{0.926 \text{ amu}}$

Atomic mass of Si $= 28.09$ amu (4 S Fs)

4

Compounds and Their Bonds

4.1 Atoms with 1, 2, or 3 valence electrons will lose those electrons to acquire a noble gas electron configuration.

   a. loses 1 e^-

   b. loses 2 e^-

   c. loses 3 e^-

   d. loses 1 e^-

   e. loses 2 e^-

4.3 **a.** The element with 3 protons is lithium. In an ion of lithium with 2 electrons, the ionic charge would be 1+, (3+) + (2–) = 1+. The lithium ion is written Li^+.

   b. The element with 9 protons is fluorine. In an ion of fluorine with 10 electrons, the ionic charge would be 1–, (9+) + (10–) = 1–. The fluoride ion is written F^-.

   c. The element with 12 protons is magnesium. In an ion of magnesium with 10 electrons, the ionic charge would be 2+, (12+) + (10–) = 2+. The magnesium ion is written Mg^{2+}.

   d. The element with 26 protons is iron. In an ion of iron with 23 electrons, the ionic charge would be 3+, (26+) + (23–) = 3+. This iron ion is written Fe^{3+}.

4.5 Atoms form ions by losing or gaining valence electrons to achieve the same electron configuration as the nearest noble gas. Elements in Groups 1A (1), 2A (2), and 3A (13) lose valence electrons, whereas elements in Groups 5A (15), 6A (16), and 7A (17) gain valence electrons to complete octets.

   a. Sr loses 2 e^-.

   b. P gains 3 e^-.

   c. Elements in Group 7A (17) gain 1 e^-.

   d. Na loses 1 e^-.

   e. Br gains 1 e^-.

4.7 **a.** Chlorine in Group 7A (17) gains 1 electron to form chloride ion Cl^-.

   b. Cesium in Group 1A (1) loses 1 electron to form cesium ion Cs^+.

   c. Sulfur in Group 6A (16) gains 2 electrons to form sulfide ion S^{2-}.

   d. Radium in Group 2A (2) loses 2 electrons to form radium ion Ra^{2+}.

4.9 **a** (Li and Cl) and **c** (K and O) will form ionic compounds.

4.11 **a.** Na^+ and $O^{2-} \rightarrow Na_2O$

   b. Al^{3+} and $Br^- \rightarrow AlBr_3$

   c. Ba^{2+} and $N^{3-} \rightarrow Ba_3N_2$

   d. Mg^{2+} and $F^- \rightarrow MgF_2$

   e. Al^{3+} and $S^{2-} \rightarrow Al_2S_3$

4.13 **a.** Ions: K^+ and $S^{2-} \rightarrow K_2S$ Check: $2(1+) + 1(2-) = 0$

b. Ions: Na^+ and $N^{3-} \rightarrow Na_3N$ Check: $3(1+) + 1(3-) = 0$

c. Ions: Al^{3+} and $I^- \rightarrow AlI_3$ Check: $1(3+) + 3(1-) = 0$

d. Ions: Ga^{3+} and $O^{2-} \rightarrow Ga_2O_3$ Check: $2(3+) + 3(2-) = 0$

4.15 **a.** aluminum oxide
b. calcium chloride
c. sodium oxide
d. magnesium phosphide
e. potassium iodide
f. barium fluoride

4.17 Most of the transition elements form more than one positive ion. The specific ion is indicated in a name by writing a Roman numeral that is the same as the ionic charge. For example, iron forms Fe^{2+} and Fe^{3+} ions, which are named iron(II) and iron(III).

4.19 **a.** iron(II)
b. copper(II)
c. zinc
d. lead(IV)
e. chromium(III)
f. manganese(II)

4.21 **a.** Ions: Sn^{2+} and $Cl^- \rightarrow$ tin(II) chloride

b. Ions: Fe^{2+} and $O^{2-} \rightarrow$ iron(II) oxide

c. Ions: Cu^+ and $S^{2-} \rightarrow$ copper(I) sulfide

d. Ions: Cu^{2+} and $S^{2-} \rightarrow$ copper(II) sulfide

e. Ions: Cd^{2+} and $Br^- \rightarrow$ cadmium bromide

f. Ions: Zn^{2+} and $Cl^- \rightarrow$ zinc chloride

4.23 **a.** Au^{3+}

b. Fe^{3+}

c. Pb^{4+}

d. Sn^{2+}

4.25 **a.** Ions: Mg^{2+} and $Cl^- \rightarrow MgCl_2$

b. Ions: Na^+ and $S^{2-} \rightarrow Na_2S$

c. Ions: Cu^+ and $O^{2-} \rightarrow Cu_2O$

d. Ions: Zn^{2+} and $P^{3-} \rightarrow Zn_3P_2$

e. Ions: Au^{3+} and $N^{3-} \rightarrow AuN$

f. Ions: Cr^{2+} and $Cl^- \rightarrow CrCl_2$

4.27 **a.** HCO_3^-

b. NH_4^+

c. PO_4^{3-}

d. HSO_4^-

4.29 **a.** sulfate
b. carbonate
c. phosphate
d. nitrate

4.31

	NO_2^-	CO_3^{2-}	HSO_4^-	PO_4^{3-}
Li^+	$LiNO_2$	Li_2CO_3	$LiHSO_4$	Li_3PO_4
Cu^{2+}	$Cu(NO_2)_2$	$CuCO_3$	$Cu(HSO_4)_2$	$Cu_3(PO_4)_2$
Ba^{2+}	$Ba(NO_2)_2$	$BaCO_3$	$Ba(HSO_4)_2$	$Ba_3(PO_4)_2$

4.33 **a.** CO_3^{2-}, sodium carbonate

b. NH_4^+, ammonium chloride

c. PO_4^{3-}, potassium phosphate

d. NO_2^-, chromium(II) nitrite

e. SO_3^{2-}, iron(II) sulfite

4.35 **a.** Ions: Ba^{2+} and $OH^- \rightarrow Ba(OH)_2$

b. Ions: Na^+ and $SO_4^{2-} \rightarrow Na_2SO_4$

c. Ions: Fe^{2+} and $NO_3^- \rightarrow Fe(NO_3)_2$

d. Ions: Zn^{2+} and $PO_4^{3-} \rightarrow Zn_3(PO_4)_2$

e. Ions: Fe^{3+} and $CO_3^{2-} \rightarrow Fe_2(CO_3)_3$

4.37 The nonmetallic elements that are not noble gases share electrons to form covalent compounds.

4.39 **a.** $1\ Br(7\ e^-) + 1\ Br(7\ e^-) = 7 + 7 = 14$ valence electrons

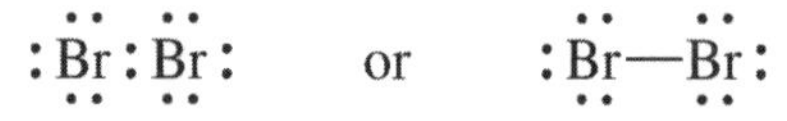

$$:\overset{..}{\underset{..}{Br}}:\overset{..}{\underset{..}{Br}}: \quad \text{or} \quad :\overset{..}{\underset{..}{Br}}—\overset{..}{\underset{..}{Br}}:$$

b. $1\ H(1\ e^-) + 1\ H(1\ e^-) = 1 + 1 = 2$ valence electrons

$$H:H \quad \text{or} \quad H—H$$

c. $1\ H(1\ e^-) + 1\ F(7\ e^-) = 1 + 7 = 8$ valence electrons

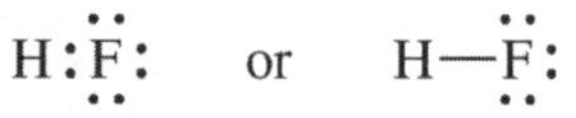

$$H:\overset{..}{\underset{..}{F}}: \quad \text{or} \quad H—\overset{..}{\underset{..}{F}}:$$

d. $1\ O(6\ e^-) + 2\ F(7\ e^-) = 6 + 14 = 20$ valence electrons

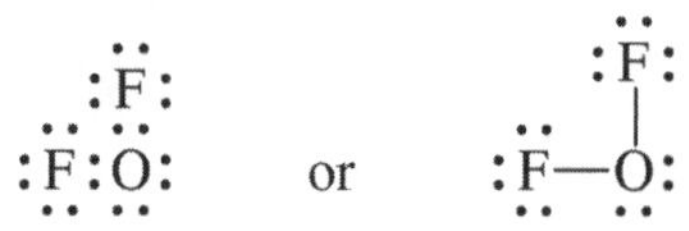

$$\overset{..}{\underset{..}{:F:}}\\ :\overset{..}{\underset{..}{F}}:\overset{..}{\underset{..}{O}}: \quad \text{or} \quad :\overset{..}{\underset{..}{F}}—\overset{..}{\underset{..}{O}}:$$

4.41 When naming covalent compounds, prefixes are used to indicate the number of each atom as shown in the subscripts of the formula. The first nonmetal is named by its elemental name; the second nonmetal is named by using its elemental name with the ending changed to *ide*.

    a. 1 P and 3 Br → phosphorus tribromide

    b. 1 C and 4 Br → carbon tetrabromide

    c. 1 Si and 2 O → silicon dioxide

    d. 1 H and 1 F → hydrogen fluoride

    e. 1 N and 3 I → nitrogen triiodide

4.43 When naming covalent compounds, prefixes are used to indicate the number of each atom as shown in the subscripts of the formula. The first nonmetal is named by its elemental name; the second nonmetal is named by using its elemental name with the ending changed to *ide*.

    a. 2 N and 3 O → dinitrogen trioxide

    b. 2 Si and 6 Br → disilicon hexabromide

    c. 4 P and 3 S → tetraphosphorus trisulfide

    d. 1 P and 5 Cl → phosphorus pentachloride

    e. 2 N and 3 S → dinitrogen trisulfide

4.45
    a. 1 C and 4 Cl → CCl_4

    b. 1 C and 1 O → CO

    c. 1 P and 3 Cl → PCl_3

    d. 2 N and 4 O → N_2O_4

4.47
    a. 1 O and 2 F → OF_2

    b. 1 B and 3 Cl → BCl_3

    c. 2 N and 3 O → N_2O_3

    d. 1 S and 6 F → SF_6

4.49
    a. Ions: Al^{3+} and SO_4^{2-} → aluminum sulfate

    b. Ions: Ca^{2+} and CO_3^{2-} → calcium carbonate

    c. 2N and 1 O → dinitrogen oxide

    d. Ions: Na^+ and PO_4^{3-} → sodium phosphate

    e. Ions: NH_4^+ and SO_4^{2-} → ammonium sulfate

    f. Ions: Fe^{3+} and O^{2-} → iron(III) oxide

4.51 The electronegativity values increase going from left to right across a period.

4.53
    a. Electronegativity increases going up a group: K, Na, Li.

    b. Electronegativity increases going left to right across a period: Na, P, Cl.

    c. Electronegativity increases going across a period and at the top of a group: Ca, Se, O.

4.55
    a. Si — Br electronegativity difference $2.8 - 1.8 = 1.0$, polar covalent

    b. Li — F electronegativity difference $4.0 - 1.0 = 3.0$, ionic

    c. Br — F electronegativity difference $4.0 - 2.8 = 1.2$, polar covalent

    d. I — I electronegativity difference $2.5 - 2.5 = 0.0$, nonpolar covalent

    e. N — P electronegativity difference $3.0 - 2.1 = 0.9$, polar covalent

    f. C — O electronegativity difference $3.5 - 2.5 = 1.0$, polar covalent

4.57 A dipole arrow points from the atom with the lower electronegativity value (more positive) to the atom in the bond that has the higher electronegativity value (more negative).

a.
$$\overset{\delta^+ \quad \delta^-}{\text{N}\!-\!\text{F}}$$
$$\longrightarrow$$

b.
$$\overset{\delta^+ \quad \delta^-}{\text{Si}\!-\!\text{Br}}$$
$$\longrightarrow$$

c.
$$\overset{\delta^+ \quad \delta^-}{\text{C}\!-\!\text{O}}$$
$$\longrightarrow$$

d.
$$\overset{\delta^+ \quad \delta^-}{\text{P}\!-\!\text{Br}}$$
$$\longrightarrow$$

e.
$$\overset{\delta^- \quad \delta^+}{\text{N}\!-\!\text{P}}$$
$$\longleftarrow$$

4.59 a. Four electron groups around a central atom have a tetrahedral electron arrangement. With two bonded atoms and two lone pairs, the shape is bent (109°).

  b. Four electron groups around a central atom have a tetrahedral electron arrangement. With three bonded atoms and one lone pair, the shape is trigonal pyramidal.

4.61 The four electron groups in PCl_3 have a tetrahedral arrangement, but three bonded atoms and one lone pair around a central atom give a trigonal pyramidal shape. The arrangement of electron pairs determines the angles between the pairs, whereas the number of bonded atoms determines the shape of the molecule.

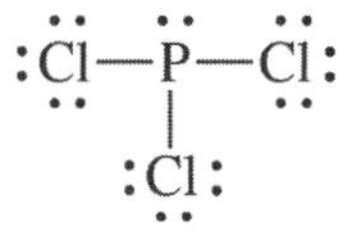

$$:\!\ddot{C}l\!-\!\overset{..}{P}\!-\!\ddot{C}l\!:$$
$$|$$
$$:\!\ddot{C}l\!:$$

4.63 In both PH_3 and NH_3, there are three bonded atoms and one lone pair on the central atoms. The shapes of both are trigonal pyramidal.

$$\text{H}\!-\!\overset{..}{\text{P}}\!-\!\text{H} \qquad \text{H}\!-\!\overset{..}{\text{N}}\!-\!\text{H}$$
$$| \qquad\qquad |$$
$$\text{H} \qquad\qquad \text{H}$$

4.65 a. The central Se atom has four electron pairs, but only two are bonded to bromine atoms. Its shape is bent (109°).

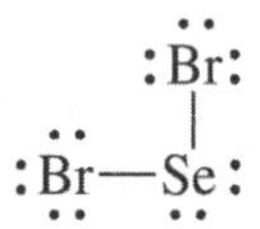

$$:\!\ddot{B}r\!:$$
$$|$$
$$:\!\ddot{B}r\!-\!\ddot{S}e\!:$$

  b. The central atom C has four electron pairs bonded to four chlorine atoms; CCl_4 has a tetrahedral shape.

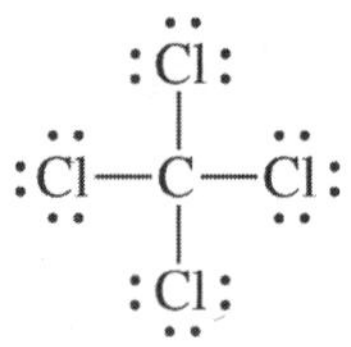

$$:\!\ddot{C}l\!:$$
$$|$$
$$:\!\ddot{C}l\!-\!C\!-\!\ddot{C}l\!:$$
$$|$$
$$:\!\ddot{C}l\!:$$

c. The central atom Ga has three electron groups bonded to three chlorine atoms; $GaCl_3$ has a trigonal planar shape.

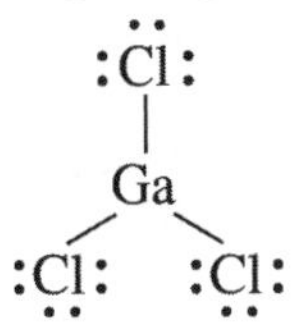

$$:\ddot{C}l:$$
$$|$$
$$Ga$$
$$:\ddot{C}l: \quad :\ddot{C}l:$$

d. The central atom Se has three electron groups with two bonded atoms and a lone pair, which gives SeO_2 a bent shape with bond angle 120°.

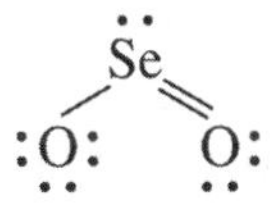

$$\ddot{S}e$$
$$:\ddot{O}: \quad \ddot{O}:$$

4.67 Cl_2 is a nonpolar molecule because there is a nonpolar covalent bond between Cl atoms, which have identical electronegativity values. In HCl, the bond is a polar bond, which makes HCl a polar molecule.

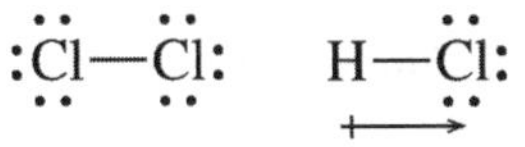

$$:\ddot{C}l—\ddot{C}l: \qquad H—\ddot{C}l:$$

nonpolar   polar

4.69 **a.** The molecule HBr contains the polar covalent H—Br bond; this single dipole makes HBr a polar molecule.

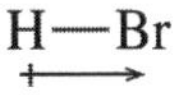

H—Br

polar

b. The molecule NF_3 contains three polar covalent N—F bonds and a lone pair on the central N atom. This asymmetric trigonal pyramidal shape makes NF_3 a polar molecule.

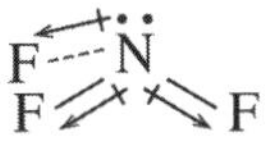

F---N
F   F

polar

c. In the molecule CHF_3, there are three polar C—F bonds and one nonpolar C C—H H bond, which makes CHF_3 a polar molecule.

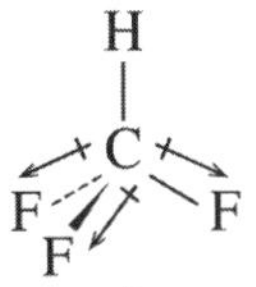

H
|
C
F   F
F

polar

4.71 **a.** BrF is a polar molecule. An attraction between the positive end of one polar molecule and the negative end of another polar molecule is called dipole–dipole attraction.
b. An ionic bond is an attraction between a positive and negative ion, as in KCl.
c. Cl_2 is a nonpolar molecule. The weak attractions that occur between temporary dipoles in nonpolar molecules are dispersion forces.
d. CH_4 is a nonpolar molecule. The weak attractions that occur between temporary dipoles in nonpolar molecules are called dispersion forces.

4.73 **a.** Hydrogen bonds are strong dipole–dipole attractions that occur between a partially positive hydrogen atom of one molecule and one of the strongly electronegative atoms F, O, or N in another, as is seen with H_2O molecules.

      b. Dispersion forces occur between temporary dipoles in Ar atoms.

      c. HBr is a polar molecule. Dipole–dipole attractions occur between dipoles in polar molecules.

      d. NF_3 is a polar molecule. Dipole–dipole attractions occur between dipoles in polar molecules.

      e. CO is a polar molecule. Dipole–dipole attractions occur between dipoles in polar molecules.

4.75 **a.** By losing one valence electron from the third energy level, sodium achieves an octet in the second energy level.

      b. The sodium ion Na^+ has the same electron arrangement as the noble gas Ne (2,8).

      c. Atoms of Group 1A (1) and 2A (2) elements can lose 1 or 2 electrons respectively to attain a noble gas electron configuration. In combination with nonmetal atoms that gain 1 or more valence electrons to achieve octet structure, they form ionic compounds. Atoms of Group 8A (18) elements already have an octet of valence electrons (two for helium), so they do not lose or gain electrons and are not normally found in compounds.

4.77 **a.** The element with 15 protons is phosphorus. In an ion of phosphorus with 18 electrons, the ionic charge would be 3–, (15+) + (18–) = 3–. The phosphide ion is written P^{3-}.

      b. The element with 8 protons is oxygen. Since there are also 8 electrons, this is an oxygen (O) atom.

      c. The element with 30 protons is zinc. In an ion of zinc with 28 electrons, the ionic charge would be 2+, (30+) + (28–) = 2+. The zinc ion is written Zn^{2+}.

      d. The element with 26 protons is iron. In an ion of iron with 23 electrons, the ionic charge would be 3+, (26+) + (23–) = 3+. This iron ion is written Fe^{3+}.

4.79 **a.** X is in Group 1A (1); Y is in Group 6A (16)

      b. ionic

      c. X^+, Y^{2-}

      d. X_2Y

      e. X^+ and $S^{2-} \rightarrow X_2S$

      f. YCl_2

      g. covalent

4.81 **a.** 2; trigonal pyramidal shape, polar molecule

      b. 1; bent shape (109°), polar molecule

      c. 3; tetrahedral shape, nonpolar molecule

4.83

Electron Arrangements		Cation	Anion	Formula of Compound	Name of Compound
2,8,2	2,5	Mg^{2+}	N^{3-}	Mg_3N_2	Magnesium nitride
2,8,8,1	2,6	K^+	O^{2-}	K_2O	Potassium oxide
2,8,3	2,8,7	Al^{3+}	Cl^-	$AlCl_3$	Aluminum chloride

4.85 **a.** 2 valence electrons, 1 bonding pair, no lone pairs

      b. 8 valence electrons, 1 bonding pair, 3 lone pairs

      c. 14 valence electrons, 1 bonding pair, 6 lone pairs

4.87 **a.** N^{3-} has an electron arrangement of 2,8, which is the same as neon (Ne).

      b. Mg^{2+} has an electron arrangement of 2,8, which is the same as neon (Ne).

      c. P^{3-} has an electron arrangement of 2,8,8, which is the same as argon (Ar).

      d. Al^{3+} has an electron arrangement of 2,8, which is the same as neon (Ne).

      e. Li^{+} has an electron arrangement of 2, which is the same as helium (He).

4.89 **a.** An element that forms an ion with a 2+ charge would be in Group 2A (2).

      b. The electron-dot symbol for an element in Group 2A (2) is $\overset{\bullet}{X}\bullet$

      c. Mg is the Group 2A (2) element in Period 3.

      d. Ions: X^{2+} and $N^{3-} \rightarrow X_3N_2$

4.91 **a.** Tin(IV) is Sn^{4+}.

      b. The Sn^{4+} ion has 50 protons and $50-4=46$ electrons.

      c. Ions: Sn^{4+} and $O^{2-} \rightarrow SnO_2$

      d. Ions: Sn^{4+} and $PO_4^{3-} \rightarrow Sn_3(PO_4)_4$

4.93 **a.** Ions: Sn^{2+} and $S^{2-} \rightarrow SnS$

      b. Ions: Pb^{4+} and $O^{2-} \rightarrow PbO_2$

      c. Ions: Ag^{+} and $Cl^{-} \rightarrow AgCl$

      d. Ions: Ca^{2+} and $N^{3-} \rightarrow Ca_3N_2$

      e. Ions: Cu^{+} and $P^{3-} \rightarrow Cu_3P$

      f. Ions: Cr^{2+} and $Br^{-} \rightarrow CrBr_2$

4.95 **a.** 1 N and 3 Cl $\rightarrow$ nitrogen trichloride

      b. 2 N and 3 S $\rightarrow$ dinitrogen trisulfide

      c. 2 N and 1 O $\rightarrow$ dinitrogen oxide

      d. 2 F $\rightarrow$ fluorine (named as the element)

      e. 1 P and 5 Cl $\rightarrow$ phosphorus pentachloride

      f. 2 P and 5 O $\rightarrow$ diphosphorus pentoxide

4.97 **a.** 1 C and 1 O $\rightarrow$ CO

      b. 2 P and 5 O $\rightarrow P_2O_5$

      c. 2 H and 1 S $\rightarrow H_2S$

      d. 1 S and 2 Cl $\rightarrow SCl_2$

4.99 **a.** Ionic, ions are Fe^{3+} and $Cl^{-} \rightarrow$ iron(III) chloride

      b. Ionic, ions are Na^{+} and $SO_4^{2-} \rightarrow$ sodium sulfate

      c. Covalent, 1 N and 2 O $\rightarrow$ nitrogen dioxide

      d. Covalent, diatomic element $\rightarrow$ nitrogen

      e. Covalent, 1 P and 5 F $\rightarrow$ phosphorus pentafluoride

      f. Covalent, 1 C and 4 F $\rightarrow$ carbon tetrafluoride

4.101 **a.** Ions: Sn^{2+} and $CO_3^{2-} \rightarrow SnCO_3$

      b. Ions: Li^+ and $P^{3-} \rightarrow Li_3P$

      c. Covalent, 1 Si and 4 Cl $\rightarrow SiCl_4$

      d. Ions: Mn^{3+} and $O^{2-} \rightarrow Mn_2O_3$

      e. Covalent, diatomic element $\rightarrow Br_2$

      f. Ions: Ca^{2+} and $Br^- \rightarrow CaBr_2$

4.103 **a.** C — O $(3.5 - 2.5 = 1.0)$ is more polar than C — N $(3.5 - 3.0 = 0.5)$.

      b. N — F $(4.0 - 3.0 = 1.0)$ is more polar than N — Br $(3.0 - 2.8 = 0.2)$.

      c. S — Cl $(3.0 - 2.5 = 0.5)$ is more polar than Br — Cl $(3.0 - 2.8 = 0.2)$.

      d. Br — I $(2.8 - 2.5 = 0.3)$ is more polar than Br — Cl $(3.0 - 2.8 = 0.2)$.

      e. N — F $(4.0 - 3.0 = 1.0)$ is more polar than N — O $(3.5 - 3.0 = 0.5)$.

4.105 **a.** Si — Cl electronegativity difference $3.0 - 1.8 = 1.2$, polar covalent

      b. C — C electronegativity difference $2.5 - 2.5 = 0.0$, nonpolar covalent

      c. Na — Cl electronegativity difference $3.0 - 0.9 = 2.1$, ionic

      d. C — H electronegativity difference $2.5 - 2.1 = 0.4$, nonpolar covalent

      e. F — F electronegativity difference $4.0 - 4.0 = 0.0$, nonpolar covalent

4.107 **a.** NH_3 is a polar molecule. The dipoles from the three polar covalent N — H bonds do not cancel because of the asymmetric trigonal pyramidal shape of the molecule.

      b. The molecule CH_3Cl has a tetrahedral shape but the atoms bonded to the central atom, C, are not identical. The three C—H bonds are nonpolar, but the C—Cl bond is polar. The nonpolar C—H bonds do not cancel the dipole from the polar C—Cl bond, making it a polar molecule.

      c. The molecule SiF_4 consists of a central atom, Si, with four polar Si — F bonds. Because it has a tetrahedral shape, the four dipoles cancel, which makes SiF_4 a nonpolar molecule.

4.109 **a.** A molecule that has a central atom with three bonded atoms and one lone pair will have a trigonal pyramidal shape. This asymmetric shape means the dipoles do not cancel, and the molecule will be polar.

      b. A molecule that has a central atom with two bonded atoms and two lone pairs will have a bent shape (109°). This asymmetric shape means the dipoles do not cancel, and the molecule will be polar.

4.111 **a.** The molecule SI_2 contains two nonpolar covalent bonds (EN $2.5 - 2.5 = 0$) and two lone pairs on the central S atom. The molecule has a bent shape (109°), but since there are no dipoles from the S—I bonds, SI_2 is a nonpolar molecule.

b. The molecule PBr_3 contains three polar covalent P—Br bonds (EN $2.8 - 2.1 = 0.7$) and a lone pair on the central P atom. This asymmetric trigonal pyramidal shape makes PBr_3 a polar molecule.

4.113 **a.** Hydrogen bonding (3) involves strong dipole–dipole attractions that occur between a partially positive hydrogen atom of one polar molecule and one of the strongly electronegative atoms F, O, or N in another, as is seen with NH_3 molecules.

b. HI is a polar molecule. Dipole–dipole attractions (2) occur between dipoles in polar molecules.

c. Dispersion forces (4) occur between temporary dipoles in nonpolar Br_2 molecules.

d. Ionic bonds (1) are strong attractions between positive and negative ions, as in Cs_2O.

4.115

Atom or Ion	Number of Protons	Number of Electrons	Electrons Lost/Gained
K^+	$19\ p^+$	$18\ e^-$	$1\ e^-$ lost
Mg^{2+}	$12\ p^+$	$10\ e^-$	$2\ e^-$ lost
O^{2-}	$8\ p^+$	$10\ e^-$	$2\ e^-$ gained
Al^{3+}	$13\ p^+$	$10\ e^-$	$3\ e^-$ lost

4.117 **a.** X as a X^{3+} ion would be in Group 3A (13).

b. X as a X^{2-} ion would be in Group 6A (16).

c. X as a X^{2+} ion would be in Group 2A (2).

4.119 Compounds with a metal and nonmetal are classified as ionic; compounds with two nonmetals are covalent.

a. Ionic, ions are Li^+ and $O^{2-} \rightarrow$ lithium oxide

b. Ionic, ions are Cr^{2+} and $NO_3^- \rightarrow$ chromium (II) nitrate

c. Ionic, ions are Mg^{2+} and $HCO_3^- \rightarrow$ magnesium bicarbonate or magnesium hydrogen carbonate

d. Covalent, 1 N and 3 F $\rightarrow$ nitrogen trifluoride

e. Ionic, ions are Ca^{2+} and $Cl^- \rightarrow$ calcium chloride

f. Ionic, ions are K^+ and $PO_4^{3-} \rightarrow$ potassium phosphate

g. Ionic, ions are Au^{3+} and $SO_3^{2-} \rightarrow$ gold(III) sulfite

h. Covalent, iodine (diatomic element)

Chemical Quantities and Reactions

5.1 One mole contains 6.02×10^{23} atoms of an element, molecules of a covalent substance, or formula units of an ionic substance.

5.3 **a.** $0.500 \text{ mole C} \times \dfrac{6.02 \times 10^{23} \text{ atoms C}}{1 \text{ mole C}} = 3.01 \times 10^{23}$ atoms of C (3 SFs)

b. $1.28 \text{ moles SO}_2 \times \dfrac{6.02 \times 10^{23} \text{ molecules SO}_2}{1 \text{ mole SO}_2} = 7.71 \times 10^{23}$ molecules of SO_2 (3 SFs)

c. $5.22 \times 10^{22} \text{ atoms Fe} \times \dfrac{1 \text{ mole Fe}}{6.02 \times 10^{23} \text{ atoms Fe}} = 0.0867$ mole of Fe (3 SFs)

d. $8.50 \times 10^{24} \text{ atoms C}_2\text{H}_6\text{O} \times \dfrac{1 \text{ mole C}_2\text{H}_6\text{O}}{6.02 \times 10^{23} \text{ atoms C}_2\text{H}_6\text{O}} = 14.1$ moles of C_2H_6O (3 SFs)

5.5 1 mole of H_3PO_4 contains 3 moles of H atoms, 1 mole of P atoms, and 4 moles of O atoms.

a. $2.00 \text{ moles H}_3\text{PO}_4 \times \dfrac{3 \text{ moles H}}{1 \text{ mole H}_3\text{PO}_4} = 6.00$ moles of H (3 SFs)

b. $2.00 \text{ moles H}_3\text{PO}_4 \times \dfrac{4 \text{ moles O}}{1 \text{ mole H}_3\text{PO}_4} = 8.00$ moles of O (3 SFs)

c. $2.00 \text{ moles H}_3\text{PO}_4 \times \dfrac{1 \text{ mole P}}{1 \text{ mole H}_3\text{PO}_4} \times \dfrac{6.02 \times 10^{23} \text{ atoms P}}{1 \text{ mole P}} = 1.20 \times 10^{24}$ atoms of P (3 SFs)

d. $2.00 \text{ moles H}_3\text{PO}_4 \times \dfrac{4 \text{ moles O}}{1 \text{ mole H}_3\text{PO}_4} \times \dfrac{6.02 \times 10^{23} \text{ atoms O}}{1 \text{ mole O}} = 4.82 \times 10^{24}$ atoms of O (3 SFs)

5.7 The subscripts indicate the moles of each element in one mole of that compound.

a. $1.0 \text{ mole quinine} \times \dfrac{24 \text{ moles H}}{1 \text{ mole quinine}} = 24$ moles of H (2 SFs)

b. $5.0 \text{ moles quinine} \times \dfrac{20 \text{ moles C}}{1 \text{ mole quinine}} = 1.0 \times 10^2$ moles of C (2 SFs)

c. $0.020 \text{ mole quinine} \times \dfrac{2 \text{ moles N}}{1 \text{ mole quinine}} = 0.040$ mole of N (2 SFs)

5.9 **a.** $1 \text{ mole K} \times \dfrac{39.1 \text{ g K}}{1 \text{ mole K}} = 39.1$ g of K

$4 \text{ moles C} \times \dfrac{12.0 \text{ g C}}{1 \text{ mole C}} = 48.0$ g of C

$5 \text{ moles H} \times \dfrac{1.01 \text{ g H}}{1 \text{ mole H}} = 5.05$ g of H

$$6 \text{ moles O} \times \frac{16.0 \text{ g O}}{1 \text{ mole O}} = 96.0 \text{ g of O}$$

1 mole of K	=	39.1 g of K
4 moles of C	=	48.0 g of C
5 moles of H	=	5.05 g of H
6 moles of O	=	96.0 g of O

$\therefore$ Molar mass of $KC_4H_5O_6$ = $\overline{188.2 \text{ g}}$

b. $2 \text{ moles Fe} \times \dfrac{55.9 \text{ g Fe}}{1 \text{ mole Fe}} = 111.8 \text{ g of Fe}$

$3 \text{ moles O} \times \dfrac{16.0 \text{ g O}}{1 \text{ mole O}} = 48.0 \text{ g of O}$

2 moles of Fe	=	111.8 g of Fe
3 moles of O	=	48.0 g of O

$\therefore$ Molar mass of Fe_2O_3 = $\overline{159.8 \text{ g}}$

c. $19 \text{ moles C} \times \dfrac{12.0 \text{ g C}}{1 \text{ mole C}} = 228.0 \text{ g of C}$

$20 \text{ moles H} \times \dfrac{1.01 \text{ g H}}{1 \text{ mole H}} = 20.2 \text{ g of H}$

$1 \text{ mole F} \times \dfrac{19.0 \text{ g F}}{1 \text{ mole F}} = 19.0 \text{ g of F}$

$1 \text{ mole N} \times \dfrac{14.0 \text{ g N}}{1 \text{ mole N}} = 14.0 \text{ g of N}$

$3 \text{ moles O} \times \dfrac{16.0 \text{ g O}}{1 \text{ mole O}} = 48.0 \text{ g of O}$

19 moles of C	=	228.0 g of C
20 moles of H	=	20.2 g of H
1 mole of F	=	19.0 g of F
1 mole of N	=	14.0 g of N
3 moles of O	=	48.0 g of O

$\therefore$ Molar mass of $C_{19}H_{20}FNO_3$ = $\overline{329.2 \text{ g}}$

d. $2 \text{ moles Al} \times \dfrac{27.0 \text{ g Al}}{1 \text{ mole Al}} = 54.0 \text{ g of Al}$

$3 \text{ moles S} \times \dfrac{32.1 \text{ g S}}{1 \text{ mole S}} = 96.3 \text{ g of S}$

$12 \text{ moles O} \times \dfrac{16.0 \text{ g O}}{1 \text{ mole O}} = 192.0 \text{ g of O}$

2 moles of Al	=	54.0 g of Al
3 moles of S	=	96.3 g of S
12 moles of O	=	192.0 g of O

$\therefore$ Molar mass of $Al_2(SO_4)_3$ = $\overline{342.3 \text{ g}}$

e. $1 \text{ mole Mg} \times \dfrac{24.3 \text{ g Mg}}{1 \text{ mole Mg}} = 24.3 \text{ g of Mg}$

$2 \text{ moles O} \times \dfrac{16.0 \text{ g O}}{1 \text{ mole O}} = 32.0 \text{ g of O}$

$2 \text{ moles H} \times \dfrac{1.01 \text{ g H}}{1 \text{ mole H}} = 2.02 \text{ g of H}$

1 mole of Mg	= 24.3 g of Mg
2 moles of O	= 32.0 g of O
2 moles of H	= 2.02 g of H
$\therefore$ Molar mass of $Mg(OH)_2$	= 58.3 g

f. $16 \text{ moles C} \times \dfrac{12.0 \text{ g C}}{1 \text{ mole C}} = 192.0 \text{ g of C}$

$19 \text{ moles H} \times \dfrac{1.01 \text{ g H}}{1 \text{ mole H}} = 19.2 \text{ g of H}$

$3 \text{ moles N} \times \dfrac{14.0 \text{ g N}}{1 \text{ mole N}} = 42.0 \text{ g of N}$

$5 \text{ moles O} \times \dfrac{16.0 \text{ g O}}{1 \text{ mole O}} = 80.0 \text{ g of O}$

$1 \text{ mole S} \times \dfrac{32.1 \text{ g S}}{1 \text{ mole S}} = 32.1 \text{ g of S}$

16 moles of C	= 192.0 g of C
19 moles of H	= 19.2 g of H
3 moles of N	= 42.0 g of N
5 moles of O	= 80.0 g of O
1 mole of S	= 32.1 g of S
$\therefore$ Molar mass of $C_{16}H_{19}N_3O_5S$	= 365.3 g

5.11 a. $2 \text{ moles Cl} \times \dfrac{35.5 \text{ g Cl}}{1 \text{ mole Cl}} = 71.0 \text{ g of Cl}$

$\therefore$ Molar mass of $Cl_2 = 71.0$ g

b. $3 \text{ moles C} \times \dfrac{12.0 \text{ g C}}{1 \text{ mole C}} = 36.0 \text{ g of C}$

$6 \text{ moles H} \times \dfrac{1.01 \text{ g H}}{1 \text{ mole H}} = 6.06 \text{ g of H}$

$3 \text{ moles O} \times \dfrac{16.0 \text{ g O}}{1 \text{ mole O}} = 48.0 \text{ g of O}$

3 moles of C	= 36.0 g of C
6 moles of H	= 6.06 g of H
3 moles of O	= 48.0 g of O
$\therefore$ Molar mass of $C_3H_6O_3$	= 90.1 g

c. $3 \text{ moles Mg} \times \dfrac{24.3 \text{ g Mg}}{1 \text{ mole Mg}} = 72.9 \text{ g of Mg}$

$2 \text{ moles P} \times \dfrac{31.0 \text{ g P}}{1 \text{ mole P}} = 62.0 \text{ g of P}$

$8 \text{ moles O} \times \dfrac{16.0 \text{ g O}}{1 \text{ mole O}} = 128.0 \text{ g of O}$

3 moles of Mg $\qquad = \quad$ 72.9 g of Mg

2 moles of P $\qquad = \quad$ 62.0 g of P

8 moles of O $\qquad = \quad$ 128.0 g of O

$\therefore$ Molar mass of $Mg_3(PO_4)_2 \quad = \quad$ 262.9 g

d. $1 \text{ mole Al} \times \dfrac{27.0 \text{ g Al}}{1 \text{ mole Al}} = 27.0 \text{ g of Al}$

$3 \text{ moles F} \times \dfrac{19.0 \text{ g F}}{1 \text{ mole F}} = 57.0 \text{ g of F}$

1 mole of Al $\qquad = \quad$ 27.0 g of Al

3 moles of F $\qquad = \quad$ 57.0 g of F

$\therefore$ Molar mass of $AlF_3 \quad = \quad$ 84.0 g

e. $2 \text{ moles C} \times \dfrac{12.0 \text{ g C}}{1 \text{ mole C}} = 24.0 \text{ g of C}$

$4 \text{ moles H} \times \dfrac{1.01 \text{ g H}}{1 \text{ mole H}} = 4.04 \text{ g of H}$

$2 \text{ moles Cl} \times \dfrac{35.5 \text{ g Cl}}{1 \text{ mole Cl}} = 71.0 \text{ g of Cl}$

2 moles of C $\qquad = \quad$ 24.0 g of C

4 moles of H $\qquad = \quad$ 4.04 g of H

2 moles of Cl $\qquad = \quad$ 71.0 g of Cl

$\therefore$ Molar mass of $C_2H_4Cl_2 = \quad$ 99.0 g

f. $1 \text{ mole Sn} \times \dfrac{118.7 \text{ g Sn}}{1 \text{ mole Sn}} = 118.7 \text{ g of Sn}$

$2 \text{ moles F} \times \dfrac{19.0 \text{ g F}}{1 \text{ mole F}} = 38.0 \text{ g of F}$

1 mole of Sn $\qquad = \quad$ 118.7 g of Sn

2 moles of F $\qquad = \quad$ 38.0 g of F

$\therefore$ Molar mass of $SnF_2 \quad = \quad$ 156.7 g

5.13 **a.** $2.00 \text{ moles Na} \times \dfrac{23.0 \text{ g Na}}{1 \text{ mole Na}} = 46.0 \text{ g of Na (3 SFs)}$

b. $2.80 \text{ moles Ca} \times \dfrac{40.1 \text{ g Ca}}{1 \text{ mole Ca}} = 112 \text{ g of Ca (3 SFs)}$

c. $0.125 \text{ mole Sn} \times \dfrac{118.7 \text{ g Sn}}{1 \text{ mole Sn}} = 14.8 \text{ g of Sn (3 SFs)}$

d. $1.76 \text{ moles Cu} \times \dfrac{63.6 \text{ g Cu}}{1 \text{ mole Cu}} = 112$ g of Cu (3 SFs)

5.15 **a.** $0.500 \text{ mole NaCl} \times \dfrac{58.5 \text{ g NaCl}}{1 \text{ mole NaCl}} = 29.3$ g of NaCl (3 SFs)

b. $1.75 \text{ moles Na}_2\text{O} \times \dfrac{62.0 \text{ g Na}_2\text{O}}{1 \text{ mole Na}_2\text{O}} = 109$ g of Na_2O (3 SFs)

c. $0.225 \text{ mole H}_2\text{O} \times \dfrac{18.0 \text{ g H}_2\text{O}}{1 \text{ mole H}_2\text{O}} = 4.05$ g of H_2O (3 SFs)

d. $4.42 \text{ moles CO}_2 \times \dfrac{44.0 \text{ g CO}_2}{1 \text{ mole CO}_2} = 194$ g of CO_2 (3 SFs)

5.17 **a.** $5.00 \text{ moles MgSO}_4 \times \dfrac{120.4 \text{ g MgSO}_4}{1 \text{ mole MgSO}_4} = 602$ g of $MgSO_4$ (3 SFs)

b. $0.25 \text{ mole CO}_2 \times \dfrac{44.0 \text{ g CO}_2}{1 \text{ mole CO}_2} = 11$ g of CO_2 (2 SFs)

5.19 **a.** $50.0 \text{ g Ag} \times \dfrac{1 \text{ mole Ag}}{107.9 \text{ g Ag}} = 0.463$ mole of Ag (3 SFs)

b. $0.200 \text{ g C} \times \dfrac{1 \text{ mole C}}{12.0 \text{ g C}} = 0.0167$ mole of C (3 SFs)

c. $15.0 \text{ g NH}_3 \times \dfrac{1 \text{ mole NH}_3}{17.0 \text{ g NH}_3} = 0.882$ mole of NH_3 (3 SFs)

d. $75.0 \text{ g SO}_2 \times \dfrac{1 \text{ mole SO}_2}{64.1 \text{ g SO}_2} = 1.17$ moles of SO_2 (3 SFs)

5.21 **a.** $25.0 \text{ g Ne} \times \dfrac{1 \text{ mole Ne}}{20.2 \text{ g Ne}} = 1.24$ moles of Ne (3 SFs)

b. $25.0 \text{ g O}_2 \times \dfrac{1 \text{ mole O}_2}{32.0 \text{ g O}_2} = 0.781$ mole of O_2 (3 SFs)

c. $25.0 \text{ g Al(OH)}_3 \times \dfrac{1 \text{ mole Al(OH)}_3}{78.0 \text{ g Al(OH)}_3} = 0.321$ mole of $Al(OH)_3$ (3 SFs)

d. $25.0 \text{ g Ga}_2\text{S}_3 \times \dfrac{1 \text{ mole Ga}_2\text{S}_3}{235.7 \text{ g Ga}_2\text{S}_3} = 0.106$ mole of Ga_2S_3 (3 SFs)

5.23 **a.** $25 \text{ g S} \times \dfrac{1 \text{ mole S}}{32.1 \text{ g S}} = 0.78$ mole of S (2 SFs)

b. $125 \text{ g SO}_2 \times \dfrac{1 \text{ mole SO}_2}{64.1 \text{ g SO}_2} \times \dfrac{1 \text{ mole S}}{1 \text{ mole SO}_2} = 1.95$ moles of S (3 SFs)

c. $2.0 \text{ moles Al}_2\text{S}_3 \times \dfrac{3 \text{ moles S}}{1 \text{ mole Al}_2\text{S}_3} = 6.0$ moles of S (2 SFs)

5.25 **a.** $1.50 \text{ moles } C_3H_8 \times \dfrac{44.1 \text{ g } C_3H_8}{1 \text{ mole } C_3H_8} = 66.2 \text{ g of } C_3H_8 \text{ (3 SFs)}$

b. $34.0 \text{ g } C_3H_8 \times \dfrac{1 \text{ mole } C_3H_8}{44.1 \text{ g } C_3H_8} = 0.771 \text{ mole of } C_3H_8 \text{ (3 SFs)}$

c. $34.0 \text{ g } C_3H_8 \times \dfrac{1 \text{ mole } C_3H_8}{44.1 \text{ g } C_3H_8} \times \dfrac{3 \text{ moles } C}{1 \text{ mole } C_3H_8} \times \dfrac{12.0 \text{ g } C}{1 \text{ mole } C} = 27.8 \text{ g of } C \text{ (3 SFs)}$

5.27 An equation is balanced when there are equal numbers of atoms of each element on the reactant side and on the product side.
   a. not balanced $2 \text{ O} \neq 3 \text{ O}$
   b. balanced
   c. not balanced $2 \text{ O} \neq 1 \text{ O}$
   d. balanced

5.29 Place coefficients in front of formulas until you make the atoms of each element equal on each side of the equation.
   a. $N_2(g) + O_2(g) \longrightarrow 2NO(g)$

   b. $2HgO(s) \longrightarrow 2Hg(l) + O_2(g)$

   c. $4Fe(s) + 3O_2(g) \longrightarrow 2Fe_2O_3(s)$

   d. $2Na(s) + Cl_2(g) \longrightarrow 2NaCl(s)$

5.31 **a.** There are two NO_3^- ions in the product. Balance by placing a 2 before $AgNO_3$.

   $Mg(s) + 2AgNO_3(aq) \longrightarrow Mg(NO_3)_2(aq) + 2Ag(s)$

   b. Start with the formula $Al_2(SO_4)_3$. Balance the Al by writing 2Al and balance the SO_4^{2-} ions by writing $3CuSO_4$.

   $2Al(s) + 3CuSO_4(aq) \longrightarrow 3Cu(s) + Al_2(SO_4)_3(aq)$

   c. $Pb(NO_3)_2(aq) + 2NaCl(aq) \longrightarrow PbCl_2(s) + 2NaNO_3(aq)$

   d. $2Al(s) + 6HCl(aq) \longrightarrow 2AlCl_3(aq) + 3H_2(g)$

5.33 **a.** This is a decomposition reaction because a single reactant splits into two simpler substances (elements).
   b. This is a single replacement reaction because one element in the reacting compound is replaced by the other reactant.

5.35 **a.** combination
   b. single replacement
   c. decomposition
   d. double replacement
   e. decomposition
   f. double replacement
   g. combination
   h. combustion

5.37 **a.** $Mg(s) + Cl_2(g) \longrightarrow MgCl_2(s)$

   b. $2HBr(g) \xrightarrow{\Delta} H_2(g) + Br_2(g)$

    c. $Mg(s) + Zn(NO_3)_2(aq) \longrightarrow Zn(s) + Mg(NO_3)_2(aq)$

    d. $K_2S(aq) + Pb(NO_3)_2(aq) \longrightarrow 2KNO_3(aq) + PbS(s)$

    e. $2C_2H_6(g) + 7O_2(g) \xrightarrow{\Delta} 4CO_2(g) + 6H_2O(g)$

5.39   Oxidation is the loss of electrons; reduction is the gain of electrons.

    a. Na^+ gains an electron to form Na; this is a reduction.

    b. Ni loses electrons to form Ni^{2+}; this is an oxidation.

    c. Cr^{3+} gains electrons to form Cr; this is a reduction.

    d. $2H^+$ gain electrons to form H_2; this is a reduction.

5.41   An oxidized substance has lost electrons; a reduced substance has gained electrons.

    a. Zn loses electrons and is oxidized. Cl_2 gains electrons and is reduced.

    b. Br^- (in $NaBr$) loses electrons and is oxidized. Cl_2 gains electrons and is reduced.

    c. O^{2-} (in PbO) loses electrons and is oxidized. Pb^{2+} (in PbO) gains electrons and is reduced.

    d. Sn^{2+} loses electrons and oxidized. Fe^{3+} gains electrons and is reduced.

5.43   **a.** Fe^{3+} gains an electron to form Fe^{2+}; this is a reduction.

    b. Fe^{2+} loses an electron to form Fe^{3+}; this is an oxidation.

5.45   Linoleic acid gains hydrogen atoms and is reduced.

5.47   **a.** $\dfrac{2 \text{ moles } SO_2}{1 \text{ mole } O_2}$ and $\dfrac{1 \text{ mole } O_2}{2 \text{ moles } SO_2}$; $\dfrac{2 \text{ moles } SO_2}{2 \text{ moles } SO_3}$ and $\dfrac{2 \text{ moles } SO_3}{2 \text{ moles } SO_2}$; $\dfrac{1 \text{ mole } SO_2}{2 \text{ moles } SO_3}$ and $\dfrac{2 \text{ moles } SO_3}{1 \text{ mole } SO_2}$

    b. $\dfrac{4 \text{ moles } P}{5 \text{ moles } O_2}$ and $\dfrac{5 \text{ moles } O_2}{4 \text{ moles } P}$; $\dfrac{4 \text{ moles } P}{2 \text{ moles } P_2O_5}$ and $\dfrac{2 \text{ moles } P_2O_5}{4 \text{ moles } P}$; $\dfrac{5 \text{ moles } O_2}{2 \text{ moles } P_2O_5}$ and $\dfrac{2 \text{ moles } P_2O_5}{5 \text{ moles } O_2}$

5.49   **a.** $2.0 \text{ moles } H_2 \times \dfrac{1 \text{ mole } O_2}{2 \text{ moles } H_2} = 1.0 \text{ mole of } O_2$ (2 SFs)

    b. $5.0 \text{ moles } O_2 \times \dfrac{2 \text{ moles } H_2}{1 \text{ mole } O_2} = 10. \text{ moles of } H_2$ (2 SFs)

    c. $2.5 \text{ moles } O_2 \times \dfrac{2 \text{ moles } H_2O}{1 \text{ mole } O_2} = 5.0 \text{ moles of } H_2O$ (2 SFs)

5.51   **a.** $0.500 \text{ mole } SO_2 \times \dfrac{5 \text{ moles } C}{2 \text{ moles } SO_2} = 1.25 \text{ moles of } C$ (3 SFs)

    b. $1.2 \text{ moles } C \times \dfrac{4 \text{ moles } CO}{5 \text{ moles } C} = 0.96 \text{ mole of } CO$ (2 SFs)

    c. $0.50 \text{ mole } CS_2 \times \dfrac{2 \text{ moles } SO_2}{1 \text{ mole } CS_2} = 1.0 \text{ mole of } SO_2$ (2 SFs)

    d. $2.5 \text{ moles } C \times \dfrac{1 \text{ mole } CS_2}{5 \text{ moles } C} = 0.50 \text{ mole of } CS_2$ (2 SFs)

5.53 **a.** $57.5 \text{ g Na} \times \dfrac{1 \text{ mole Na}}{23.0 \text{ g Na}} \times \dfrac{2 \text{ moles Na}_2\text{O}}{4 \text{ moles Na}} \times \dfrac{62.0 \text{ g Na}_2\text{O}}{1 \text{ mole Na}_2\text{O}} = 77.5 \text{ g of Na}_2\text{O (3 SFs)}$

b. $18.0 \text{ g Na} \times \dfrac{1 \text{ mole Na}}{23.0 \text{ g Na}} \times \dfrac{1 \text{ mole O}_2}{4 \text{ moles Na}} \times \dfrac{32.0 \text{ g O}_2}{1 \text{ mole O}_2} = 6.26 \text{ g of O}_2 \text{ (3 SFs)}$

c. $75.0 \text{ g Na}_2\text{O} \times \dfrac{1 \text{ mole Na}_2\text{O}}{62.0 \text{ g Na}_2\text{O}} \times \dfrac{1 \text{ mole O}_2}{2 \text{ moles Na}_2\text{O}} \times \dfrac{32.0 \text{ g O}_2}{1 \text{ mole O}_2} = 19.4 \text{ g of O}_2 \text{ (3 SFs)}$

5.55 **a.** $13.6 \text{ g NH}_3 \times \dfrac{1 \text{ mole NH}_3}{17.0 \text{ g NH}_3} \times \dfrac{3 \text{ moles O}_2}{4 \text{ moles NH}_3} \times \dfrac{32.0 \text{ g O}_2}{1 \text{ mole O}_2} = 19.2 \text{ g of O}_2 \text{ (3 SFs)}$

b. $6.50 \text{ g O}_2 \times \dfrac{1 \text{ mole O}_2}{32.0 \text{ g O}_2} \times \dfrac{2 \text{ moles N}_2}{3 \text{ moles O}_2} \times \dfrac{28.0 \text{ g N}_2}{1 \text{ mole N}_2} = 3.79 \text{ g of N}_2 \text{ (3 SFs)}$

c. $34.0 \text{ g NH}_3 \times \dfrac{1 \text{ mole NH}_3}{17.0 \text{ g NH}_3} \times \dfrac{6 \text{ moles H}_2\text{O}}{4 \text{ moles NH}_3} \times \dfrac{18.0 \text{ g H}_2\text{O}}{1 \text{ mole H}_2\text{O}} = 54.0 \text{ g of H}_2\text{O (3 SFs)}$

5.57 **a.** $28.0 \text{ g NO}_2 \times \dfrac{1 \text{ mole NO}_2}{46.0 \text{ g NO}_2} \times \dfrac{1 \text{ mole H}_2\text{O}}{3 \text{ moles NO}_2} \times \dfrac{18.0 \text{ g H}_2\text{O}}{1 \text{ mole H}_2\text{O}} = 3.65 \text{ g of H}_2\text{O (3 SFs)}$

b. $15.8 \text{ g NO}_2 \times \dfrac{1 \text{ mole NO}_2}{46.0 \text{ g NO}_2} \times \dfrac{1 \text{ mole NO}}{3 \text{ moles NO}_2} \times \dfrac{30.0 \text{ g NO}}{1 \text{ mole NO}} = 3.43 \text{ g of NO (3 SFs)}$

c. $8.25 \text{ g NO}_2 \times \dfrac{1 \text{ mole NO}_2}{46.0 \text{ g NO}_2} \times \dfrac{2 \text{ moles HNO}_3}{3 \text{ moles NO}_2} \times \dfrac{63.0 \text{ g HNO}_3}{1 \text{ mole HNO}_3} = 7.53 \text{ g of HNO}_3 \text{ (3 SFs)}$

5.59 **a.** $2\text{PbS}(s) + 3\text{O}_2(g) \xrightarrow{\Delta} 2\text{PbO}(s) + 2\text{SO}_2(g)$

b. $0.125 \text{ mole PbS} \times \dfrac{3 \text{ moles O}_2}{2 \text{ moles PbS}} \times \dfrac{32.0 \text{ g O}_2}{1 \text{ mole O}_2} = 6.00 \text{ g of O}_2 \text{ (3 SFs)}$

c. $65.0 \text{ g PbS} \times \dfrac{1 \text{ mole PbS}}{239.3 \text{ g PbS}} \times \dfrac{2 \text{ moles SO}_2}{2 \text{ moles PbS}} \times \dfrac{64.1 \text{ g SO}_2}{1 \text{ mole SO}_2} = 17.4 \text{ g of SO}_2 \text{ (3 SFs)}$

d. $128 \text{ g PbO} \times \dfrac{1 \text{ mole PbO}}{223.2 \text{ g PbO}} \times \dfrac{2 \text{ moles PbS}}{2 \text{ moles PbO}} \times \dfrac{239.3 \text{ g PbS}}{1 \text{ mole PbS}} = 137 \text{ g of PbS (3 SFs)}$

5.61 **a.** The energy of activation is the energy required to break the bonds of the reacting molecules.
b. A catalyst provides a pathway that lowers the activation energy and speeds up a reaction.
c. In exothermic reactions, the energy of the products is lower than the energy of the reactants.

d.

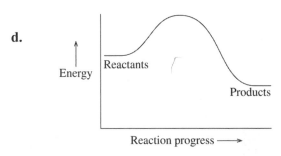

5.63 **a.** An exothermic reaction releases heat.
  b. An endothermic reaction has a lower energy level for the reactants than the products.
  c. The metabolism of glucose is an exothermic reaction, providing energy for the body.

5.65 **a.** Heat is a product; the reaction is exothermic.
  b. Heat is a reactant; the reaction is endothermic.
  c. Heat is a product; the reaction is exothermic.

5.67 **a.** The rate of a reaction tells how fast the products are formed or how fast the reactants are consumed.
  b. Because more reactants will have the energy necessary to proceed to products (the activation energy) at room temperature than at refrigerator temperatures, the rate of formation of mold will be higher at room temperature.

5.69 **a.** Addition of a reactant increases the reaction rate.
  b. Increasing the temperature increases the number of collisions with the energy of activation. The rate of reaction will be increased.
  c. Addition of a catalyst increases the reaction rate.
  d. Removal of reactant decreases the reaction rate.

5.71 **1.** **a.** S_2Cl_2
  b. Molar mass of $S_2Cl_2 = 2(32.1 \text{ g}) + 2(35.5 \text{ g}) = 135.2 \text{ g}$ (4 SFs)
  c. $10.0 \text{ g } S_2Cl_2 \times \dfrac{1 \text{ mole } S_2Cl_2}{135.2 \text{ g } S_2Cl_2} = 0.0740 \text{ mole of } S_2Cl_2$ (3 SFs)

  2. **a.** C_6H_6
  b. Molar mass of $C_6H_6 = 6(12.0 \text{ g}) + 6(1.01 \text{ g}) = 78.1 \text{ g}$ (3 SFs)
  c. $10.0 \text{ g } C_6H_6 \times \dfrac{1 \text{ mole } C_6H_6}{78.1 \text{ g } C_6H_6} = 0.128 \text{ mole of } C_6H_6$ (3 SFs)

5.73 **a.** Molar mass of dipyrithione ($C_{10}H_8N_2O_2S_2$)
  $= 10(12.0 \text{ g}) + 8(1.01 \text{ g}) + 2(14.0 \text{ g}) + 2(16.0 \text{ g}) + 2(32.1 \text{ g}) = 252.3 \text{ g}$ (4 SFs)
  b. $25.0 \text{ g } C_{10}H_8N_2O_2S_2 \times \dfrac{1 \text{ mole } C_{10}H_8N_2O_2S_2}{252.3 \text{ g } C_{10}H_8N_2O_2S_2} = 0.0991 \text{ mole of } C_{10}H_8N_2O_2S_2$ (3 SFs)
  c. $25.0 \text{ g } C_{10}H_8N_2O_2S_2 \times \dfrac{1 \text{ mole } C_{10}H_8N_2O_2S_2}{252.3 \text{ g } C_{10}H_8N_2O_2S_2} \times \dfrac{10 \text{ moles C}}{1 \text{ mole } C_{10}H_8N_2O_2S_2}$
  $= 0.991 \text{ mole of C}$ (3 SFs)
  d. $8.2 \times 10^{24} \text{ N atoms} \times \dfrac{1 \text{ mole N}}{6.02 \times 10^{23} \text{ N atoms}} \times \dfrac{1 \text{ mole } C_{10}H_8O_2N_2S_2}{2 \text{ moles N}}$
  $= 6.8 \text{ moles of } C_{10}H_8O_2N_2S_2$ (2 SFs)

5.75 **a.** 1, 1, 2 combination
  b. 2, 2, 1 decomposition

5.77 **a.** $2NO(g) + O_2(g) \rightarrow 2NO_2(g)$
  b. combination

5.79 **a.** $2NI_3(g) \rightarrow N_2(g) + 3I_2(g)$

   b. decomposition

5.81 **a.** $2Cl_2(g) + O_2(g) \rightarrow 2OCl_2(g)$

   b. combination

5.83 **a.** $1 \text{ mole Zn} \times \dfrac{65.4 \text{ g Zn}}{1 \text{ mole Zn}} = 65.4 \text{ g of Zn}$

   $1 \text{ mole S} \times \dfrac{32.1 \text{ g S}}{1 \text{ mole S}} = 32.1 \text{ g of S}$

   $4 \text{ moles O} \times \dfrac{16.0 \text{ g O}}{1 \text{ mole O}} = 64.0 \text{ g of O}$

1 mole of Zn	=	65.4 g of Zn
1 mole of S	=	32.1 g of S
4 moles of O	=	64.0 g of O
$\therefore$ Molar mass of $ZnSO_4$ =		161.5 g

   b. $1 \text{ mole Ca} \times \dfrac{40.1 \text{ g Ca}}{1 \text{ mole Ca}} = 40.1 \text{ g of Ca}$

   $2 \text{ moles I} \times \dfrac{126.9 \text{ g I}}{1 \text{ mole I}} = 253.8 \text{ g of I}$

   $6 \text{ moles O} \times \dfrac{16.0 \text{ g O}}{1 \text{ mole O}} = 96.0 \text{ g of O}$

1 mole of Ca	=	40.1 g of Ca
2 moles of I	=	253.8 g of I
6 moles of O	=	96.0 g of O
$\therefore$ Molar mass of $Ca(IO_3)_2$ =		389.9 g

   c. $5 \text{ moles C} \times \dfrac{12.0 \text{ g C}}{1 \text{ mole C}} = 60.0 \text{ g of C}$

   $8 \text{ moles H} \times \dfrac{1.01 \text{ g H}}{1 \text{ mole H}} = 8.08 \text{ g of H}$

   $1 \text{ mole N} \times \dfrac{14.0 \text{ g N}}{1 \text{ mole N}} = 14.0 \text{ g of N}$

   $1 \text{ mole Na} \times \dfrac{23.0 \text{ g Na}}{1 \text{ mole Na}} = 23.0 \text{ g of Na}$

   $4 \text{ moles O} \times \dfrac{16.0 \text{ g O}}{1 \text{ mole O}} = 64.0 \text{ g of O}$

$$5 \text{ moles of C} \qquad = \quad 60.0 \text{ g of C}$$
$$8 \text{ moles of H} \qquad = \quad 8.08 \text{ g of H}$$
$$1 \text{ mole of N} \qquad = \quad 14.0 \text{ g of N}$$
$$1 \text{ mole of Na} \qquad = \quad 23.0 \text{ g of Na}$$
$$4 \text{ moles of O} \qquad = \quad 64.0 \text{ g of O}$$
$$\therefore \text{ Molar mass of } C_5H_8NNaO_4 = \overline{169.1 \text{ g}}$$

d. $6 \text{ moles C} \times \dfrac{12.0 \text{ g C}}{1 \text{ mole C}} = 72.0 \text{ g of C}$

$12 \text{ moles H} \times \dfrac{1.01 \text{ g H}}{1 \text{ mole H}} = 12.1 \text{ g of H}$

$2 \text{ moles O} \times \dfrac{16.0 \text{ g O}}{1 \text{ mole O}} = 32.0 \text{ g of O}$

$$6 \text{ moles of C} \qquad = \quad 72.0 \text{ g of C}$$
$$12 \text{ moles of H} \qquad = \quad 12.1 \text{ g of H}$$
$$2 \text{ moles of O} \qquad = \quad 32.0 \text{ g of O}$$
$$\therefore \text{ Molar mass of } C_6H_{12}O_2 = \overline{116.1 \text{ g}}$$

5.85 **a.** $0.150 \text{ mole K} \times \dfrac{39.1 \text{ g K}}{1 \text{ mole K}} = 5.87 \text{ g of K (3 SFs)}$

b. $0.150 \text{ mole } Cl_2 \times \dfrac{71.0 \text{ g } Cl_2}{1 \text{ mole } Cl_2} = 10.7 \text{ g of } Cl_2 \text{ (3 SFs)}$

c. $0.150 \text{ mole } Na_2CO_3 \times \dfrac{106.0 \text{ g } Na_2CO_3}{1 \text{ mole } Na_2CO_3} = 15.9 \text{ g of } Na_2CO_3 \text{ (3 SFs)}$

5.87 **a.** $25.0 \text{ g } CO_2 \times \dfrac{1 \text{ mole } CO_2}{44.0 \text{ g } CO_2} = 0.568 \text{ mole of } CO_2 \text{ (3 SFs)}$

b. $25.0 \text{ g } Al_2O_3 \times \dfrac{1 \text{ mole } Al_2O_3}{102.0 \text{ g } Al_2O_3} = 0.245 \text{ mole of } Al_2O_3 \text{ (3 SFs)}$

c. $25.0 \text{ g } MgCl_2 \times \dfrac{1 \text{ mole } MgCl_2}{95.3 \text{ g } MgCl_2} = 0.262 \text{ mole of } MgCl_2 \text{ (3 SFs)}$

5.89 **a.** Atoms of a metal and a nonmetal forming an ionic compound is a combination reaction.
   b. When a compound of hydrogen and carbon reacts with oxygen, it is a combustion reaction.
   c. When calcium carbonate is heated to produce calcium oxide and carbon dioxide, it is a decomposition reaction.
   d. Zinc replacing copper in $Cu(NO_3)_2$ is a single replacement reaction.

5.91 **a.** $NH_3(g) + HCl(g) \rightarrow NH_4Cl(s)$ combination

   b. $C_5H_{12}(g) + 8O_2(g) \xrightarrow{\Delta} 5\,CO_2(g) + 6H_2O(g)$ combustion

   c. $2Sb(s) + 3Cl_2(g) \rightarrow 2SbCl_3(s)$ combination

   d. $2NI_3(s) \rightarrow N_2(g) + 3I_2(g)$ decomposition

    e. $2KBr(aq) + Cl_2(aq) \rightarrow 2KCl(aq) + Br_2(l)$ single replacement

    f. $2Fe(s) + 3H_2SO_4(aq) \rightarrow Fe_2(SO_4)_3(aq) + 3H_2(g)$ single replacement

    g. $Al_2(SO_4)_3(aq) + 6NaOH(aq) \rightarrow 3Na_2SO_4(aq) + 2Al(OH)_3(s)$ double replacement

5.93   **a.** $Zn(s) + 2HCl(aq) \rightarrow ZnCl_2(aq) + H_2(g)$

       b. $BaCO_3(s) \xrightarrow{\Delta} BaO(s) + CO_2(g)$

       c. $NaOH(aq) + HCl(aq) \rightarrow NaCl(aq) + H_2O(l)$

       d. $2Al(s) + 3F_2(g) \rightarrow 2AlF_3(s)$

5.95   **a.** Zn^{2+} gains electrons to form Zn; this is a reduction.

       b. Al loses electrons to form Al^{3+}; this is an oxidation.

       c. Pb loses electrons to form Pb^{2+}; this is an oxidation.

       d. Cl_2 gains electrons to form $2Cl^-$; this is a reduction.

5.97   **a.** $2NH_3(g) + 5F_2(g) \rightarrow N_2F_4(g) + 6HF(g)$

       b. $4.00 \text{ moles HF} \times \dfrac{2 \text{ moles } NH_3}{6 \text{ moles HF}} = 1.33$ moles of NH_3 (3 SFs)

           $4.00 \text{ moles HF} \times \dfrac{5 \text{ moles } F_2}{6 \text{ moles HF}} = 3.33$ moles of F_2 (3 SFs)

       c. $25.5 \text{ g } NH_3 \times \dfrac{1 \text{ mole } NH_3}{17.0 \text{ g } NH_3} \times \dfrac{5 \text{ moles } F_2}{2 \text{ moles } NH_3} \times \dfrac{38.0 \text{ g } F_2}{1 \text{ mole } F_2} = 143$ g of F_2 (3 SFs)

       d. $3.40 \text{ g } NH_3 \times \dfrac{1 \text{ mole } NH_3}{17.0 \text{ g } NH_3} \times \dfrac{1 \text{ mole } N_2F_4}{2 \text{ moles } NH_3} \times \dfrac{104.0 \text{ g } N_2F_4}{1 \text{ mole } N_2F_4} = 10.4$ g of N_2F_4 (3 SFs)

5.99   **a.** $C_5H_{12}(g) + 8O_2(g) \xrightarrow{\Delta} 5CO_2(g) + 6H_2O(g) + \text{energy}$

       b. $72 \text{ g } H_2O \times \dfrac{1 \text{ mole } H_2O}{18.0 \text{ g } H_2O} \times \dfrac{1 \text{ mole } C_5H_{12}}{6 \text{ moles } H_2O} \times \dfrac{72.1 \text{ g } C_5H_{12}}{1 \text{ mole } C_5H_{12}} = 48$ g of C_5H_{12} (2 SFs)

       c. $32.0 \text{ g } O_2 \times \dfrac{1 \text{ mole } O_2}{32.0 \text{ g } O_2} \times \dfrac{5 \text{ moles } CO_2}{8 \text{ moles } O_2} \times \dfrac{44.0 \text{ g } CO_2}{1 \text{ mole } CO_2} = 27.5$ g of CO_2 (3 SFs)

5.101 **a.**   Since heat is produced during the formation of $SiCl_4$, this is an exothermic reaction.

       b.   Heat is given off; the energy of the products is lower than the energy of the reactants.

5.103 **a.** $124 \text{ g } C_2H_6O \times \dfrac{1 \text{ mole } C_2H_6O}{46.1 \text{ g } C_2H_6O} \times \dfrac{1 \text{ mole } C_6H_{12}O_6}{2 \text{ moles } C_2H_6O} \times \dfrac{180.1 \text{ g } C_6H_{12}O_6}{1 \text{ mole } C_6H_{12}O_6}$

           $= 242$ g of $C_6H_{12}O_6$ (3 SFs)

       b. $0.240 \text{ kg } C_6H_{12}O_6 \times \dfrac{1000 \text{ g}}{1 \text{ kg}} \times \dfrac{1 \text{ mole } C_6H_{12}O_6}{180.1 \text{ g } C_6H_{12}O_6} \times \dfrac{2 \text{ moles } C_2H_6O}{1 \text{ mole } C_6H_{12}O_6} \times \dfrac{46.1 \text{ g } C_2H_6O}{1 \text{ mole } C_2H_6O}$

           $= 123$ g of C_2H_6O (3 SFs)

5.105 **a.** $4Al(s) + 3O_2(g) \rightarrow 2Al_2O_3(s)$

   b. This is a combination reaction.

   c. $4.50 \text{ moles Al} \times \dfrac{3 \text{ moles } O_2}{4 \text{ moles Al}} = 3.38 \text{ moles of } O_2$ (3 SFs)

   d. $50.2 \text{ g Al} \times \dfrac{1 \text{ mole Al}}{27.0 \text{ g Al}} \times \dfrac{2 \text{ moles } Al_2O_3}{4 \text{ moles Al}} \times \dfrac{102.0 \text{ g } Al_2O_3}{1 \text{ mole } Al_2O_3} = 94.8 \text{ g of } Al_2O_3$ (3 SFs)

   e. $8.00 \text{ g } O_2 \times \dfrac{1 \text{ mole } O_2}{32.0 \text{ g } O_2} \times \dfrac{2 \text{ moles } Al_2O_3}{3 \text{ moles } O_2} \times \dfrac{102.0 \text{ g } Al_2O_3}{1 \text{ mole } Al_2O_3} = 17.0 \text{ g of } Al_2O_3$ (3 SFs)

5.107 **a.** $0.500 \text{ mole } C_3H_6O_3 \times \dfrac{6.02\times10^{23} \text{ molecules}}{1 \text{ mole } C_3H_6O_3} = 3.01\times10^{23} \text{ molecules of } C_3H_6O_3$ (3 SFs)

   b. $1.50 \text{ moles } C_3H_6O_3 \times \dfrac{3 \text{ moles C}}{1 \text{ mole } C_3H_6O_3} \times \dfrac{6.02\times10^{23} \text{ atoms}}{1 \text{ mole C}} = 2.71\times10^{24} \text{ atoms of C}$ (3 SFs)

   c. $4.5\times10^{24} \text{ O atoms} \times \dfrac{1 \text{ mole O}}{6.02\times10^{23} \text{ O atoms}} \times \dfrac{1 \text{ mole } C_3H_6O_3}{3 \text{ moles O}} = 2.5 \text{ moles of } C_3H_6O_3$ (2 SFs)

   d. Molar mass of lactic acid $(C_3H_6O_3)$
      $= 3(12.0 \text{ g}) + 6(1.01 \text{ g}) + 3(16.0 \text{ g}) = 90.1 \text{ g}$ (3 SFs)

5.109 **a.** $2C_2H_2(g) + 5O_2(g) \xrightarrow{\Delta} 4CO_2(g) + 2H_2O(g) + 2600 \text{ kJ}$

   b. Since heat is produced during the combustion of acetylene, the reaction is exothermic.

   c. $64.0 \text{ g } O_2 \times \dfrac{1 \text{ mole } O_2}{32.0 \text{ g } O_2} \times \dfrac{2 \text{ moles } H_2O}{5 \text{ moles } O_2} = 0.800 \text{ mole of } H_2O$ (3 SFs)

   d. $2.25\times10^{24} \text{ molecules } C_2H_2 \times \dfrac{1 \text{ mole } C_2H_2}{6.02\times10^{23} \text{ molecules } C_2H_2} \times \dfrac{5 \text{ moles } O_2}{2 \text{ moles } C_2H_2}$

      $= 9.34 \text{ moles of } O_2$ (3 SFs)

Answers to Combining Ideas from Chapters 3 to 5

CI.7 **a.** Y has the higher electronegativity, since it is a nonmetal in Group 7A (17).

      b. X^{2+}, Y^-

      c. X has an electron arrangement of 2,8,2.   Y has an electron arrangement of 2,8,7.

      d. X^{2+} has an electron arrangement of 2,8. Y^- has an electron arrangement of 2,8,8.

      e. X^{2+} has the same electron arrangement as the noble gas neon (Ne).

         Y^- has the same electron arrangement as the noble gas argon (Ar).

      f. $MgCl_2$, magnesium chloride

CI.9 **a.** Reactants: A and B_2; Products: AB_3:

      b. $2A + 3B_2 \rightarrow 2AB_3$

      c. combination

CI.11 **a.** The molecular formula of acetone is C_3H_6O.

      b. Molar mass of acetone $(C_3H_6O) = 3(12.0\text{ g}) + 6(1.01\text{ g}) + 1(16.0\text{ g}) = 58.1\text{ g (3 SFs)}$

      c. Polar covalent bonds in C_3H_6O: C—O

         Nonpolar covalent bonds in C_3H_6O: C—C, C—H

      d. $C_3H_6O(l) + 4O_2(g) \xrightarrow{\Delta} 3CO_2(g) + 3H_2O(g) + \text{energy}$

      e. $15.0\text{ mL } C_3H_6O \times \dfrac{0.786\text{ g } C_3H_6O}{1\text{ mL } C_3H_6O} \times \dfrac{1\text{ mole } C_3H_6O}{58.1\text{ g } C_3H_6O} \times \dfrac{4\text{ moles } O_2}{1\text{ mole } C_3H_6O} \times \dfrac{32.0\text{ g } O_2}{1\text{ mole } O_2}$

         $= 26.0\text{ g of } O_2 \text{ (3 SFs)}$

CI.13 **a.** The formula of shikimic acid is $C_7H_{10}O_5$.

      b. Molar mass of shikimic acid $(C_7H_{10}O_5) = 7(12.0\text{ g}) + 10(1.01\text{ g}) + 5(16.0\text{ g})$

         $= 174.1\text{ g (4 SFs)}$

      c. $130\text{ g shikimic acid} \times \dfrac{1\text{ mole shikimic acid}}{174.1\text{ g shikimic acid}} = 0.75\text{ mole of shikimic acid (2 SFs)}$

      d. $155\text{ g anise} \times \dfrac{0.13\text{ g shikimic acid}}{2.6\text{ g anise}} \times \dfrac{1\text{ capsule Tamiflu}}{0.13\text{ g shikimic acid}}$

         $= 59\text{ capsules of Tamiflu (2 SFs)}$

      e. Molar mass of Tamiflu $(C_{16}H_{28}N_2O_4) = 16(12.0\text{ g}) + 28(1.01\text{ g}) + 2(14.0\text{ g}) + 4(16.0\text{ g})$

         $= 312.3\text{ g (4 SFs)}$

      f. $500\,000\text{ people} \times \dfrac{2\text{ capsules}}{1\text{ day 1 person}} \times 5\text{ days} \times \dfrac{75\text{ mg Tamiflu}}{1\text{ capsule}} \times \dfrac{1\text{ g Tamiflu}}{1000\text{ mg Tamiflu}}$

         $\times \dfrac{1\text{ kg Tamiflu}}{1000\text{ g Tamiflu}} = 400\text{ kg of Tamiflu (1 SF)}$

6.1   **a.**   At higher temperatures, gas particles have greater kinetic energy, which makes them move faster.

      b.   Because there are great distances between the particles of a gas, they can be pushed closer together and still remain a gas.

6.3   **a.**   The temperature of a gas can be expressed in kelvins.

      b.   The volume of a gas can be expressed in milliliters.

      c.   The amount of a gas can be expressed in grams.

      d.   Pressure can be expressed in millimeters of mercury (mmHg).

6.5   Some units used to describe the pressure of a gas are atmospheres (abbreviated atm), mmHg, torr, pounds per square inch (lb / in.^{2} or psi), pascals, kilopascals, and in. Hg.

6.7   **a.** $2.00 \text{ atm} \times \dfrac{760 \text{ torr}}{1 \text{ atm}} = 1520 \text{ torr (3 SFs)}$

      b. $2.00 \text{ atm} \times \dfrac{760 \text{ mmHg}}{1 \text{ atm}} = 1520 \text{ mmHg (3 SFs)}$

6.9   As a diver ascends to the surface, external pressure decreases. If the air in the lungs were not exhaled, its volume would expand and severely damage the lungs. The pressure in the lungs must adjust to changes in the external pressure.

6.11   **a.**   The pressure is greater in cylinder A. According to Boyle's law, a decrease in volume pushes the gas particles closer together, which will cause an increase in the pressure.

      b.

Property	Conditions 1	Conditions 2	Know	Predict
Pressure (P)	$P_1 = 650$ mmHg	$P_2 = 1.2$ atm	P increases	
Volume (V)	$V_1 = 220$ mL	$V_2 = ?$ mL		V decreases

According to Boyle's law, $P_1V_1 = P_2V_2$, then

$$V_2 = V_1 \times \frac{P_1}{P_2} = 220 \text{ mL} \times \frac{650 \text{ mmHg}}{1.2 \text{ atm}} \times \frac{1 \text{ atm}}{760 \text{ mmHg}} = 160 \text{ mL (2 SFs)}$$

6.13   **a.**   The pressure of the gas doubles when the volume is halved.

      b.   The pressure falls to one-third the initial pressure when the volume expands to three times its initial value.

      c.   The pressure increases to ten times the original pressure when the volume decreases to one-tenth of the initial volume.

6.15   From Boyle's law, we know that pressure is inversely related to volume (e.g., pressure increases when volume decreases).

      a.   Volume increases; pressure must decrease.

$$P_2 = P_1 \times \frac{V_1}{V_2} = 655 \text{ mmHg} \times \frac{10.0 \text{ L}}{20.0 \text{ L}} = 328 \text{ mmHg (3 SFs)}$$

b. Volume decreases; pressure must increase.

$$P_2 = P_1 \times \frac{V_1}{V_2} = 655 \text{ mmHg} \times \frac{10.0 \text{ L}}{2.50 \text{ L}} = 2620 \text{ mmHg (3 SFs)}$$

c. The mL units must be converted to L for unit cancellation in the calculation, and because the volume decreases, pressure must increase.

$$P_2 = P_1 \times \frac{V_1}{V_2} = 655 \text{ mmHg} \times \frac{10.0 \text{ L}}{1500 \text{ mL}} \times \frac{1000 \text{ mL}}{1 \text{ L}} = 4370 \text{ mmHg (3 SFs)}$$

6.17 From Boyle's law, we know that pressure is inversely related to volume.

a. Pressure increases; volume must decrease.

$$V_2 = V_1 \times \frac{P_1}{P_2} = 50.0 \text{ L} \times \frac{760. \text{ mmHg}}{1500 \text{ mmHg}} = 25 \text{ L (2 SFs)}$$

b. The mmHg units must be converted to atm for unit cancellation in the calculation, and because the pressure increases, volume must decrease.

$$P_1 = 760. \text{ mmHg} \times \frac{1 \text{ atm}}{760 \text{ mmHg}} = 1.00 \text{ atm}$$

$$V_2 = V_1 \times \frac{P_1}{P_2} = 50.0 \text{ L} \times \frac{1.00 \text{ atm}}{4.0 \text{ atm}} = 12.5 \text{ L (3 SFs)}$$

c. The mmHg units must be converted to atm for unit cancellation in the calculation, and because the pressure decreases, volume must increase.

$$P_1 = 760. \text{ mmHg} \times \frac{1 \text{ atm}}{760 \text{ mmHg}} = 1.00 \text{ atm}$$

$$V_2 = V_1 \times \frac{P_1}{P_2} = 50.0 \text{ L} \times \frac{1.00 \text{ atm}}{0.500 \text{ atm}} = 100. \text{ L (3 SFs)}$$

6.19 Pressure decreases; volume must increase.

$$V_2 = V_1 \times \frac{P_1}{P_2} = 5.0 \text{ L} \times \frac{5.0 \text{ atm}}{1.0 \text{ atm}} = 25 \text{ L of cyclopropane (2 SFs)}$$

6.21 **a.** Inspiration begins when the diaphragm flattens, causing the lungs to expand. The increased volume of the thoracic cavity reduces the pressure in the lungs such that air flows into the lungs.
b. Expiration occurs as the diaphragm relaxes, causing a decrease in the volume of the lungs. The pressure of the air in the lungs increases and air flows out of the lungs.
c. Inspiration occurs when the pressure within the lungs is less than the pressure of the air in the atmosphere.

6.23 According to Charles's law, there is a direct relationship between Kelvin temperature and volume (e.g., volume increases when temperature increases, if the pressure and amount of gas remain constant).
a. Diagram C shows an increased volume corresponding to an increase in temperature.
b. Diagram A shows a decreased volume corresponding to a decrease in temperature.
c. Diagram B shows no change in volume, which corresponds to no net change in temperature.

6.25 According to Charles's law, a change in the volume of a gas is directly proportional to the change in its Kelvin temperature. In all gas law computations, temperatures must be in kelvins. (Temperatures in °C are converted to K by the addition of 273.) The initial temperature for all cases here is $T_1 = 15\ °C + 273 = 288\ K$.

a. Volume increases; temperature must have increased.

$$T_2 = T_1 \times \frac{V_2}{V_1} = 288\ K \times \frac{5.00\ \cancel{L}}{2.50\ \cancel{L}} = 576\ K \qquad 576\ K - 273 = 303\ °C\ (3\ SFs)$$

b. Volume decreases; temperature must have decreased.

$$T_2 = T_1 \times \frac{V_2}{V_1} = 288\ K \times \frac{1250\ \cancel{mL}}{2.50\ \cancel{L}} \times \frac{1\ \cancel{L}}{1000\ \cancel{mL}} = 144\ K$$

$$144\ K - 273 = -129\ °C\ (3\ SFs)$$

c. Volume increases; temperature must have increased.

$$T_2 = T_1 \times \frac{V_2}{V_1} = 288\ K \times \frac{7.50\ \cancel{L}}{2.50\ \cancel{L}} = 864\ K \qquad 864\ K - 273 = 591\ °C\ (3\ SFs)$$

d. Volume increases; temperature must have increased.

$$T_2 = T_1 \times \frac{V_2}{V_1} = 288\ K \times \frac{3550\ \cancel{mL}}{2.50\ \cancel{L}} \times \frac{1\ \cancel{L}}{1000\ \cancel{mL}} = 409\ K \qquad 409\ K - 273 = 136\ °C\ (3\ SFs)$$

6.27 According to Charles's law, a change in the volume of a gas is directly proportional to the change in its Kelvin temperature. In all gas law computations, temperatures must be in kelvins. (Temperatures in °C are converted to K by the addition of 273.) The initial temperature for all cases here is $T_1 = 75\ °C + 273 = 348\ K$.

a. When temperature decreases, volume must also decrease.

$$T_2 = 55\ °C + 273 = 328\ K$$

$$V_2 = V_1 \times \frac{T_2}{T_1} = 2500\ mL \times \frac{328\ \cancel{K}}{348\ \cancel{K}} = 2400\ mL\ (2\ SFs)$$

b. When temperature increases, volume must also increase.

$$V_2 = V_1 \times \frac{T_2}{T_1} = 2500\ mL \times \frac{680.\ \cancel{K}}{348\ \cancel{K}} = 4900\ mL\ (2\ SFs)$$

c. When temperature decreases, volume must also decrease.

$$T_2 = -25\ °C + 273 = 248\ K$$

$$V_2 = V_1 \times \frac{T_2}{T_1} = 2500\ mL \times \frac{248\ \cancel{K}}{348\ \cancel{K}} = 1800\ mL\ (2\ SFs)$$

d. When temperature decreases, volume must also decrease.

$$V_2 = V_1 \times \frac{T_2}{T_1} = 2500\ mL \times \frac{240.\ \cancel{K}}{348\ \cancel{K}} = 1700\ mL\ (2\ SFs)$$

6.29 According to Gay-Lussac's law, temperature is directly related to pressure. For example, temperature increases when the pressure increases. In all gas law computations, temperatures must be in kelvins. (Temperatures in °C are converted to K by the addition of 273.)

a. $T_1 = 25 \,°\text{C} + 273 = 298 \text{ K}$

$$T_2 = T_1 \times \frac{P_2}{P_1} = 298 \text{ K} \times \frac{620 \text{ mmHg}}{740 \text{ mmHg}} = 250. \text{ K} \qquad 250. \text{ K} - 273 = -23 \,°\text{C (2 SFs)}$$

b. $T_1 = -18 \,°\text{C} + 273 = 255 \text{ K}$

$$T_2 = T_1 \times \frac{P_2}{P_1} = 255 \text{ K} \times \frac{1250 \text{ torr}}{0.950 \text{ atm}} \times \frac{1 \text{ atm}}{760 \text{ torr}} = 441 \text{ K} \qquad 441 \text{ K} - 273 = 168 \,°\text{C (3 SFs)}$$

6.31 According to Gay-Lussac's law, temperature is directly related to pressure. For example, temperature increases when the pressure increases. In all gas law computations, temperatures must be in kelvins. (Temperatures in °C are converted to K by the addition of 273.)

a. $T_1 = 155 \,°\text{C} + 273 = 428 \text{ K} \qquad T_2 = 0 \,°\text{C} + 273 = 273 \text{ K}$

$$P_2 = P_1 \times \frac{T_2}{T_1} = 1200 \text{ torr} \times \frac{273 \text{ K}}{428 \text{ K}} = 770 \text{ torr (2 SFs)}$$

b. $T_1 = 12 \,°\text{C} + 273 = 285 \text{ K} \qquad T_2 = 35 \,°\text{C} + 273 = 308 \text{ K}$

$$P_2 = P_1 \times \frac{T_2}{T_1} = 1.40 \text{ atm} \times \frac{308 \text{ K}}{285 \text{ K}} \times \frac{760 \text{ torr}}{1 \text{ atm}} = 1150 \text{ torr (3 SFs)}$$

6.33 a. On the top of a mountain, water boils below 100 °C because the atmospheric (external) pressure is less than 1 atm. The boiling point is the temperature at which the vapor pressure of a liquid becomes equal to the external (in this case, atmospheric) pressure.

b. Because the pressure inside a pressure cooker is greater than 1 atm, water boils above 100 °C. The higher temperature of the boiling water allows food to cook more quickly.

6.35 $T_1 = 25 \,°\text{C} + 273 = 298 \text{ K}; \qquad V_1 = 6.50 \text{ L}; \qquad P_1 = 845 \text{ mmHg}$

a. $T_2 = 325 \text{ K}; \quad V_2 = 1850 \text{ mL} = 1.85 \text{ L}; \quad P_2 = ? \text{ atm}$

$$P_2 = P_1 \times \frac{V_1}{V_2} \times \frac{T_2}{T_1} = 845 \text{ mmHg} \times \frac{6.50 \text{ L}}{1.85 \text{ L}} \times \frac{325 \text{ K}}{298 \text{ K}} \times \frac{1 \text{ atm}}{760 \text{ mmHg}} = 4.26 \text{ atm (3 SFs)}$$

b. $T_2 = 12 \,°\text{C} + 273 = 285 \text{ K}; \qquad V_2 = 2.25 \text{ L}; \qquad P_2 = ? \text{ atm}$

$$P_2 = P_1 \times \frac{V_1}{V_2} \times \frac{T_2}{T_1} = 845 \text{ mmHg} \times \frac{6.50 \text{ L}}{2.25 \text{ L}} \times \frac{285 \text{ K}}{298 \text{ K}} \times \frac{1 \text{ atm}}{760 \text{ mmHg}} = 3.07 \text{ atm (3 SFs)}$$

c. $T_2 = 47 \,°\text{C} + 273 = 320. \text{ K}; \qquad V_2 = 12.8 \text{ L}; \qquad P_2 = ? \text{ atm}$

$$P_2 = P_1 \times \frac{V_1}{V_2} \times \frac{T_2}{T_1} = 845 \text{ mmHg} \times \frac{6.50 \text{ L}}{12.8 \text{ L}} \times \frac{320. \text{ K}}{298 \text{ K}} \times \frac{1 \text{ atm}}{760 \text{ mmHg}} = 0.606 \text{ atm (3 SFs)}$$

6.37 $T_1 = 225\ °C + 273 = 498\ K;$ $\qquad V_1 = 100.0\ mL;$ $\qquad P_1 = 1.80\ atm$

$T_2 = -25\ °C + 273 = 248\ K;$ $\qquad V_2 = ?\ mL;$ $\qquad P_2 = 0.80\ atm$

$$V_2 = V_1 \times \frac{P_1}{P_2} \times \frac{T_2}{T_1} = 100.0\ mL \times \frac{1.80\ atm}{0.80\ atm} \times \frac{248\ K}{498\ K} = 110\ mL\ (2\ SFs)$$

6.39 The volume increases because the number of gas particles in the tire or basketball is increased.

6.41 According to Avogadro's law, a change in the volume of a gas is directly proportional to the change in the number of moles of gas. $n_1 = 1.50$ moles of Ne; $V_1 = 8.00$ L

a. $V_2 = V_1 \times \dfrac{n_2}{n_1} = 8.00\ L \times \dfrac{\frac{1}{2}(1.50)\ moles\ Ne}{1.50\ moles\ Ne} = 4.00\ L\ (3\ SFs)$

b. $25.0\ g\ Ne \times \dfrac{1\ mole\ Ne}{20.2\ g\ Ne} = 1.24$ moles of Ne added

$n_2 = 1.50$ moles Ne $+ 1.24$ moles Ne $= 2.74$ moles of Ne

$V_2 = V_1 \times \dfrac{n_2}{n_1} = 8.00\ L \times \dfrac{2.74\ moles\ Ne}{1.50\ moles\ Ne} = 14.6\ L\ (3\ SFs)$

c. $n_2 = 1.50$ moles Ne $+ 3.50$ moles $O_2 = 5.00$ moles of gases

$V_2 = V_1 \times \dfrac{n_2}{n_1} = 8.00\ L \times \dfrac{5.00\ moles}{1.50\ moles} = 26.7\ L\ (3\ SFs)$

6.43 At STP, 1 mole of any gas occupies a volume of 22.4 L.

a. $44.8\ L\ O_2\ (STP) \times \dfrac{1\ mole\ O_2}{22.4\ L\ O_2\ (STP)} = 2.00$ moles of O_2 (3 SFs)

b. $4.00\ L\ CO_2\ (STP) \times \dfrac{1\ mole\ CO_2}{22.4\ L\ CO_2\ (STP)} = 0.179$ mole of CO_2 (3 SFs)

6.45 **a.** $6.40\ g\ O_2 \times \dfrac{1\ mole\ O_2}{32.0\ g\ O_2} \times \dfrac{22.4\ L\ O_2\ (STP)}{1\ mole\ O_2} = 4.48\ L$ of O_2 at STP (3 SFs)

b. $50.0\ g\ Ne \times \dfrac{1\ mole\ Ne}{20.2\ g\ Ne} \times \dfrac{22.4\ L\ Ne\ (STP)}{1\ mole\ Ne} \times \dfrac{1000\ mL}{1\ L} = 55\,400\ mL$ of Ne at STP (3 SFs)

6.47 In a gas mixture, the pressure that each gas exerts as part of the total pressure is called the partial pressure of that gas. Because the air sample is a mixture of gases, the total pressure is the sum of the partial pressures of each gas in the sample.

6.49 To obtain the total pressure in a gas mixture, add up all of the partial pressures using the same pressure unit.

$P_{total} = P_{Nitrogen} + P_{Oxygen} + P_{Helium} = 425\ torr + 115\ torr + 225\ torr = 765\ torr$ (3 SFs)

6.51 Because the total pressure in a gas mixture is the sum of the partial pressures using the same pressure unit, addition and subtraction are used to obtain the "missing" partial pressure.

$$P_{\text{Nitrogen}} = P_{\text{total}} - (P_{\text{Oxygen}} + P_{\text{Helium}})$$

$$= 925 \text{ torr} - (425 \text{ torr} + 75 \text{ torr}) = 425 \text{ torr (3 SFs)}$$

6.53 **a.** If oxygen cannot readily cross from the lungs into the bloodstream, then the partial pressure of oxygen will be lower in the blood of an emphysema patient.
   b. Breathing a higher concentration of oxygen will help to increase the supply of oxygen in the lungs and blood and raise the partial pressure of oxygen in the blood.

6.55 **a.** 2; the fewest number of gas particles will exert the lowest pressure.
   b. 1; the greatest number of gas particles will exert the highest pressure.

6.57 **a.** A; volume decreases when temperature decreases.
   b. C; volume increases when pressure decreases.
   c. A; volume decreases when the number of moles of gas decreases.
   d. B; doubling the Kelvin temperature would double the volume, but when half of the gas escapes, the volume would decrease by half. These two opposing effects cancel each other, and there is no overall change in the volume.
   e. C; increasing the moles of gas causes an increase in the volume to keep T and P constant.

6.59 **a.** The volume of the chest and lungs will decrease when compressed during the Heimlich maneuver.
   b. A decrease in volume causes the pressure to increase. A piece of food would be dislodged with a sufficiently high pressure.

6.61 $T_1 = -8\ ^{\circ}\text{C} + 273 = 265 \text{ K};$ $P_1 = 658 \text{ mmHg};$ $V_1 = 31\,000 \text{ L}$

   $T_2 = 0\ ^{\circ}\text{C} + 273 = 273 \text{ K};$ $P_2 = 760 \text{ mmHg};$ $V_2 = ? \text{ L}$

$$V_2 = V_1 \times \frac{P_1}{P_2} \times \frac{T_2}{T_1} = 31\,000 \text{ L} \times \frac{658 \text{ mmHg}}{760 \text{ mmHg}} \times \frac{273 \text{ K}}{265 \text{ K}} = 28\,000 \text{ L at STP}$$

$$28\,000 \text{ L H}_2\text{ (STP)} \times \frac{1 \text{ mole H}_2}{22.4 \text{ L H}_2\text{ (STP)}} \times \frac{2.02 \text{ g H}_2}{1 \text{ mole H}_2} \times \frac{1 \text{ kg H}_2}{1000 \text{ g H}_2} = 2.5 \text{ kg of H}_2\text{ (2 SFs)}$$

6.63 $T_1 = 127\ ^{\circ}\text{C} + 273 = 400.\text{ K};$ $P_1 = 2.00 \text{ atm}$
   $T_2 = ?\ ^{\circ}\text{C};$ $P_2 = 0.25 \text{ atm}$

$$T_2 = T_1 \times \frac{P_2}{P_1} = 400.\text{ K} \times \frac{0.25 \text{ atm}}{2.00 \text{ atm}} = 50.\text{ K} \qquad 50.\text{ K} - 273 = -223\ ^{\circ}\text{C}$$

6.65 **a.** $25.0 \text{ L He (STP)} \times \dfrac{1 \text{ mole He}}{22.4 \text{ L He (STP)}} \times \dfrac{4.00 \text{ g He}}{1 \text{ mole He}} = 4.46 \text{ g of He (3 SFs)}$

   b. $T_1 = 0\ ^{\circ}\text{C} + 273 = 273 \text{ K};$ $V_1 = 25.0 \text{ L};$ $P_1 = 1.00 \text{ atm}$

   $T_2 = -35\ ^{\circ}\text{C} + 273 = 238 \text{ K};$ $V_2 = 2460 \text{ L};$ $P_2 = ? \text{ mmHg}$

$$P_2 = P_1 \times \frac{V_1}{V_2} \times \frac{T_2}{T_1} = 1.00 \text{ atm} \times \frac{25.0 \text{ L}}{2460 \text{ L}} \times \frac{238 \text{ K}}{273 \text{ K}} \times \frac{760 \text{ mmHg}}{1 \text{ atm}} = 6.73 \text{ mmHg (3 SFs)}$$

6.67 Because the partial pressure of nitrogen is to be reported in torr, the atm and mmHg units (for oxygen and argon, respectively) must be converted to torr, as follows:

$$P_{\text{Oxygen}} = 0.60 \text{ atm} \times \frac{760 \text{ torr}}{1 \text{ atm}} = 460 \text{ torr} \quad \text{and} \quad P_{\text{Argon}} = 425 \text{ mmHg} \times \frac{1 \text{ torr}}{1 \text{ mmHg}} = 425 \text{ torr}$$

$$\therefore P_{\text{Nitrogen}} = P_{\text{total}} - (P_{\text{Oxygen}} + P_{\text{Argon}})$$

$$= 1250 \text{ torr} - (460 \text{ torr} + 425 \text{ torr}) = 370 \text{ torr (2 SFs)}$$

6.69 $T_1 = 24 \text{ °C} + 273 = 297 \text{ K};$ $\quad P_1 = 745 \text{ mmHg};$ $\quad V_1 = 425 \text{ mL}$

$T_2 = -95 \text{ °C} + 273 = 178 \text{ K};$ $\quad P_2 = 0.115 \text{ atm};$ $\quad V_2 = ? \text{ mL}$

$$V_2 = V_1 \times \frac{P_1}{P_2} \times \frac{T_2}{T_1} = 425 \text{ mL} \times \frac{745 \text{ mmHg}}{0.115 \text{ atm}} \times \frac{1 \text{ atm}}{760 \text{ mmHg}} \times \frac{178 \text{ K}}{297 \text{ K}} = 2170 \text{ mL (3 SFs)}$$

6.71 **a.** False. The flask containing helium gas contains more atoms because 1 gram of helium contains more moles of helium and therefore more atoms than 1 gram of neon.

b. False. There are different numbers of moles of gas in the flasks, which means that the pressures are different.

c. True. There are more moles of helium, which makes the pressure of helium greater than that of neon.

d. True. Density is mass divided by volume. If gases have the same mass and volume, they have the same density.

6.73 $T_1 = 15 \text{ °C} + 273 = 288 \text{ K};$ $\quad P_1 = 745 \text{ mmHg};$ $\quad V_1 = 4250 \text{ mL}$

$T_2 = ? \text{ °C};$ $\quad P_2 = 1.20 \text{ atm} \times \dfrac{760 \text{ mmHg}}{1 \text{ atm}} = 912 \text{ mmHg};$ $\quad V_2 = 2.50 \text{ L} \times \dfrac{1000 \text{ mL}}{1 \text{ L}} = 2.50 \times 10^3 \text{ mL}$

$$T_2 = T_1 \times \frac{V_2}{V_1} \times \frac{P_2}{P_1} = 288 \text{ K} \times \frac{2.50 \times 10^3 \text{ mL}}{4250 \text{ mL}} \times \frac{912 \text{ mmHg}}{745 \text{ mmHg}} = 207 \text{ K} - 273 = -66 \text{ °C (2 SFs)}$$

6.75 **a.** 2000: $780 \text{ Tg CO}_2 \times \dfrac{10^{12} \text{ g CO}_2}{1 \text{ Tg CO}_2} \times \dfrac{1 \text{ kg CO}_2}{1000 \text{ g CO}_2} = 7.8 \times 10^{11} \text{ kg of CO}_2 \text{ (2 SFs)}$

2020: $990 \text{ Tg CO}_2 \times \dfrac{10^{12} \text{ g CO}_2}{1 \text{ Tg CO}_2} \times \dfrac{1 \text{ kg CO}_2}{1000 \text{ g CO}_2} = 9.9 \times 10^{11} \text{ kg of CO}_2 \text{ (2 SFs)}$

b. 2000: $780 \text{ Tg CO}_2 \times \dfrac{10^{12} \text{ g CO}_2}{1 \text{ Tg CO}_2} \times \dfrac{1 \text{ mole CO}_2}{44.0 \text{ g CO}_2} = 1.8 \times 10^{13} \text{ moles of CO}_2 \text{ (2 SFs)}$

2020: $990 \text{ Tg CO}_2 \times \dfrac{10^{12} \text{ g CO}_2}{1 \text{ Tg CO}_2} \times \dfrac{1 \text{ mole CO}_2}{44.0 \text{ g CO}_2} = 2.3 \times 10^{13} \text{ moles of CO}_2 \text{ (2 SFs)}$

c. increase is 990 Tg − 780 Tg = 210 Tg

$$210 \text{ Tg CO}_2 \times \frac{10^{12} \text{ g CO}_2}{1 \text{ Tg CO}_2} \times \frac{1 \text{ Mg CO}_2}{10^6 \text{ g CO}_2} = 2.1 \times 10^8 \text{ Mg of CO}_2 \text{ increase (2 SFs)}$$

7

Solutions

7.1   The component present in the smaller amount is the solute; the larger amount is the solvent.
  a.   NaCl, solute; water, solvent
  b.   water, solute; ethanol, solvent
  c.   oxygen, solute; nitrogen, solvent

7.3   The K^+ and I^- ions at the surface of the solid are pulled into solution by the polar water molecules, where the hydration process surrounds separate ions with water molecules.

7.5   **a.**   Sodium nitrate, $NaNO_3$ (an ionic solute), would be soluble in water (a polar solvent).

  b.   Iodine, I_2 (a nonpolar solute), would be soluble in CCl_4 (a nonpolar solvent).

  c.   Sucrose (a polar solute) would be soluble in water (a polar solvent).

  d.   Gasoline (a nonpolar solute) would be soluble in CCl_4 (a nonpolar solvent).

7.7   The strong electrolyte KF completely dissociates into K^+ and F^- ions when it dissolves in water.

  When the weak electrolyte HF dissolves in water, there are a few ions of H^+ and F^- present, but mostly dissolved HF molecules.

7.9   Strong electrolytes dissociate completely into ions.

  a. $KCl(s) \xrightarrow{H_2O} K^+(aq) + Cl^-(aq)$

  b. $CaCl_2(s) \xrightarrow{H_2O} Ca^{2+}(aq) + 2Cl^-(aq)$

  c. $K_3PO_4(s) \xrightarrow{H_2O} 3K^+(aq) + PO_4^{3-}(aq)$

  d. $Fe(NO_3)_3(s) \xrightarrow{H_2O} Fe^{3+}(aq) + 3NO_3^-(aq)$

7.11   **a.** $HC_2H_3O_2(l) \underset{H_2O}{\rightleftharpoons} H^+(aq) + C_2H_3O_2^-(aq)$ mostly molecules, a few ions

  An aqueous solution of a weak electrolyte like acetic acid will contain mostly $HC_2H_3O_2$ molecules with a few H^+ ions and a few $C_2H_3O_2^-$ ions.

  b. $NaBr(s) \xrightarrow{H_2O} Na^+(aq) + Br^-(aq)$ ions only

  An aqueous solution of a strong electrolyte like NaBr will contain only the ions Na^+ and Br^-.

  c. $C_6H_{12}O_6(s) \xrightarrow{H_2O} C_6H_{12}O_6(aq)$ molecules only

  An aqueous solution of a nonelectrolyte like fructose will contain only $C_6H_{12}O_6$ molecules.

7.13   **a.** K_2SO_4 is a strong electrolyte because only ions are present in the K_2SO_4 solution.

  b. NH_4OH is a weak electrolyte because only a few NH_4^+ and OH^- ions are present in the solution.

  c. $C_6H_{12}O_6$ is a nonelectrolyte because only $C_6H_{12}O_6$ molecules are present in the solution.

7.15 **a.** $1 \text{ mole } K^+ \times \dfrac{1 \text{ Eq } K^+}{1 \text{ mole } K^+} = 1 \text{ Eq of } K^+$

b. $2 \text{ moles } OH^- \times \dfrac{1 \text{ Eq } OH^-}{1 \text{ mole } OH^-} = 2 \text{ Eq of } OH^-$

c. $1 \text{ mole } Ca^{2+} \times \dfrac{2 \text{ Eq } Ca^{2+}}{1 \text{ mole } Ca^{2+}} = 2 \text{ Eq of } Ca^{2+}$

d. $3 \text{ moles } CO_3^{2-} \times \dfrac{2 \text{ Eq } CO_3^{2-}}{1 \text{ mole } CO_3^{2-}} = 6 \text{ Eq of } CO_3^{2-}$

7.17 $1.00 \text{ L} \times \dfrac{154 \text{ mEq}}{1 \text{ L}} \times \dfrac{1 \text{ Eq}}{1000 \text{ mEq}} \times \dfrac{1 \text{ mole } Na^+}{1 \text{ Eq}} = 0.154 \text{ mole of } Na^+ (3 \text{ SFs})$

$1.00 \text{ L} \times \dfrac{154 \text{ mEq}}{1 \text{ L}} \times \dfrac{1 \text{ Eq}}{1000 \text{ mEq}} \times \dfrac{1 \text{ mole } Cl^-}{1 \text{ Eq}} = 0.154 \text{ mole of } Cl^- (3 \text{ SFs})$

7.19 In any solution, the total equivalents of anions must be equal to the equivalents of cations.

mEq of anions $= 40.$ mEq/L $Cl^- + 15$ mEq/L $HPO_4^{2-} = 55$ mEq/L of anions

mEq of $Na^+ =$ mEq of anions $= 55$ mEq/L of Na^+

7.21 **a.** The solution must be saturated because no additional solute dissolves.
       b. The solution was unsaturated because the sugar cube dissolves completely.

7.23 **a.** At 20 °C, KCl has a solubility of 34 g of KCl in 100 g of H_2O. Because 25 g of KCl is less than the maximum amount that can dissolve in 100 g of H_2O at 20 °C, the KCl solution is unsaturated.
       b. At 20 °C, $NaNO_3$ has a solubility of 88 g of $NaNO_3$ in 100 g of H_2O. Using the solubility as a conversion factor, we can calculate the maximum amount of $NaNO_3$ that can dissolve in 25 g of H_2O:

$$25 \text{ g } H_2O \times \dfrac{88 \text{ g } NaNO_3}{100 \text{ g } H_2O} = 22 \text{ g of } NaNO_3 (2 \text{ SFs})$$

Because 11 g of $NaNO_3$ is less than the maximum amount that can dissolve in 25 g of H_2O at 20 °C, the $NaNO_3$ solution is unsaturated.

c. At 20 °C, sugar has a solubility of 204 g of $C_{12}H_{22}O_{11}$ in 100 g of H_2O. Using the solubility as a conversion factor, we can calculate the maximum amount of sugar that can dissolve in 125 g of H_2O:

$$125 \text{ g } H_2O \times \dfrac{204 \text{ g sugar}}{100 \text{ g } H_2O} = 255 \text{ g of sugar } (3 \text{ SFs})$$

Because 400. g of $C_{12}H_{22}O_{11}$ exceeds the maximum amount that can dissolve in 125 g of H_2O at 20 °C, the sugar solution is saturated, and excess undissolved sugar will be present on the bottom of the container.

7.25   **a.**   At 20 °C, KCl has a solubility of 34 g of KCl in 100 g of H_2O.

$\therefore$ 200. g of H_2O will dissolve:

$$200. \text{ g } \cancel{H_2O} \times \frac{34 \text{ g KCl}}{100 \text{ g } \cancel{H_2O}} = 68 \text{ g of KCl (2 SFs)}$$

At 20 °C, 68 g of KCl will remain in solution.

b.   Since 80. g of KCl dissolves at 50 °C and 68 g remains in solution at 20 °C, the mass of solid KCl that crystallizes after cooling is (80. g KCl $-$ 68 g KCl) = 12 g of KCl (2 SFs)

7.27   **a.**   In general, the solubility of solid solutes (like sugar) increases as temperature is increased.
   b.   The solubility of a gaseous solute (CO_2) is less at a higher temperature.
   c.   The solubility of a gaseous solute is less at a higher temperature, and the CO_2 pressure in the can is increased. When the can of warm soda is opened, more CO_2 is released, producing more spray.

7.29   **a.**   Salts containing Li^+ ions are soluble.

b.   The Cl^- salt containing Ag^+ is insoluble.

c.   Salts containing CO_3^{2-} ions are usually insoluble.

d.   Salts containing K^+ ions are soluble.

e.   Salts containing NO_3^- ions are soluble.

7.31   A 5% (m/m) glucose solution contains 5 g of glucose in 100 g of solution (5 g of glucose + 95 g of water), whereas a 5% (m/v) glucose solution contains 5 g of glucose in 100 mL of solution.

7.33   **a.**   mass of solution = 25 g of KCl + 125 g of H_2O = 150. g of solution

$$\frac{25 \text{ g KCl}}{150. \text{ g solution}} \times 100\% = 17\% \text{ (m/m) KCl solution (2 SFs)}$$

b. $$\frac{12 \text{ g sugar}}{225 \text{ g solution}} \times 100\% = 5.3\% \text{ (m/m) sugar solution (2 SFs)}$$

c. $$\frac{8.0 \text{ g CaCl}_2}{80.0 \text{ g solution}} \times 100\% = 10.\% \text{ (m/m) CaCl}_2 \text{ solution (2 SFs)}$$

7.35 $$\frac{\text{mass (in grams) solute}}{\text{volume (in mL) solution}} \times 100\% = \% \text{ mass/volume}$$

a. $$\frac{75 \text{ g Na}_2\text{SO}_4}{250 \text{ mL solution}} \times 100\% = 30.\% \text{ (m/v) Na}_2\text{SO}_4 \text{ solution (2 SFs)}$$

b. $$\frac{39 \text{ g sucrose}}{355 \text{ mL solution}} \times 100\% = 11\% \text{ (m/v) sucrose solution (2 SFs)}$$

7.37   **a.** $$50.0 \text{ } \cancel{\text{mL solution}} \times \frac{5.0 \text{ g KCl}}{100 \text{ } \cancel{\text{mL solution}}} = 2.5 \text{ g of KCl (2 SFs)}$$

b. $$1250 \text{ } \cancel{\text{mL solution}} \times \frac{4.0 \text{ g NH}_4\text{Cl}}{100 \text{ } \cancel{\text{mL solution}}} = 50. \text{ g of NH}_4\text{Cl (2 SFs)}$$

c. $$250. \text{ } \cancel{\text{mL solution}} \times \frac{10.0 \text{ mL acetic acid}}{100 \text{ } \cancel{\text{mL solution}}} = 25.0 \text{ mL of acetic acid (3 SFs)}$$

7.39 $355 \text{ mL solution} \times \dfrac{22.5 \text{ mL alcohol}}{100 \text{ mL solution}} = 79.9 \text{ mL of alcohol (3 SFs)}$

7.41   **a.** $1 \text{ h} \times \dfrac{100 \text{ mL solution}}{1 \text{ h}} \times \dfrac{20. \text{ g mannitol}}{100. \text{ mL solution}} = 20. \text{ g of mannitol (2 SFs)}$

   b. $12 \text{ h} \times \dfrac{100 \text{ mL solution}}{1 \text{ h}} \times \dfrac{20. \text{ g mannitol}}{100. \text{ mL solution}} = 240 \text{ g of mannitol (2 SFs)}$

7.43 $100. \text{ g glucose} \times \dfrac{100 \text{ mL solution}}{5 \text{ g glucose}} \times \dfrac{1 \text{ L}}{1000 \text{ mL}} = 2 \text{ L of glucose solution (1 SF)}$

7.45   **a.** $5.0 \text{ g LiNO}_3 \times \dfrac{100 \text{ g solution}}{25 \text{ g LiNO}_3} = 20. \text{ g of LiNO}_3 \text{ solution (2 SFs)}$

   b. $40.0 \text{ g KOH} \times \dfrac{100 \text{ mL solution}}{10.0 \text{ g KOH}} = 400. \text{ mL of KOH solution (3 SFs)}$

   c. $2.0 \text{ mL formic acid} \times \dfrac{100 \text{ mL solution}}{10.0 \text{ mL formic acid}} = 20. \text{ mL of formic acid solution (2 SFs)}$

7.47 $\text{molarity (M)} = \dfrac{\text{moles of solute}}{\text{liters of solution}}$

   a. $\dfrac{2.00 \text{ moles glucose}}{4.00 \text{ L solution}} = 0.500 \text{ M glucose solution (3 SFs)}$

   b. $\dfrac{4.00 \text{ g KOH}}{2.00 \text{ L solution}} \times \dfrac{1 \text{ mole KOH}}{56.1 \text{ g KOH}} = 0.0357 \text{ M KOH solution (3 SFs)}$

   c. $\dfrac{5.85 \text{ g NaCl}}{400. \text{ mL solution}} \times \dfrac{1 \text{ mole NaCl}}{58.5 \text{ g NaCl}} \times \dfrac{1000 \text{ mL solution}}{1 \text{ L solution}} = 0.250 \text{ M NaCl solution (3 SFs)}$

7.49   **a.** $2.00 \text{ L solution} \times \dfrac{1.50 \text{ moles NaOH}}{1 \text{ L solution}} \times \dfrac{40.0 \text{ g NaOH}}{1 \text{ mole NaOH}} = 120. \text{ g of NaOH (3 SFs)}$

   b. $4.00 \text{ L solution} \times \dfrac{0.200 \text{ mole KCl}}{1 \text{ L solution}} \times \dfrac{74.6 \text{ g KCl}}{1 \text{ mole KCl}} = 59.7 \text{ g of KCl (3 SFs)}$

   c. $25.0 \text{ mL solution} \times \dfrac{1 \text{ L solution}}{1000 \text{ mL solution}} \times \dfrac{6.00 \text{ moles HCl}}{1 \text{ L solution}} \times \dfrac{36.5 \text{ g HCl}}{1 \text{ mole HCl}} = 5.48 \text{ g of HCl (3 SFs)}$

7.51   **a.** $3.00 \text{ moles KBr} \times \dfrac{1 \text{ L solution}}{2.00 \text{ moles KBr}} = 1.50 \text{ L of solution (3 SFs)}$

   b. $15.0 \text{ moles NaCl} \times \dfrac{1 \text{ L solution}}{1.50 \text{ moles NaCl}} = 10.0 \text{ L of solution (3 SFs)}$

   c. $0.0500 \text{ mole Ca(NO}_3)_2 \times \dfrac{1 \text{ L solution}}{0.800 \text{ mole Ca(NO}_3)_2} \times \dfrac{1000 \text{ mL solution}}{1 \text{ L solution}}$

   $= 62.5 \text{ mL of solution (3 SFs)}$

7.53   Adding water (solvent) to the soup increases the volume and dilutes the tomato soup concentration.

7.55 $C_1V_1 = C_2V_2$ or $M_1V_1 = M_2V_2$

    a. $M_2 = M_1 \times \dfrac{V_1}{V_2} = 6.0\ M \times \dfrac{2.0\ \cancel{L}}{6.0\ \cancel{L}} = 2.0\ M$ HCl solution (2 SFs)

    b. $M_2 = M_1 \times \dfrac{V_1}{V_2} = 12\ M \times \dfrac{0.50\ \cancel{L}}{3.0\ \cancel{L}} = 2.0\ M$ NaOH solution (2 SFs)

    c. $C_2 = C_1 \times \dfrac{V_1}{V_2} = 25\% \times \dfrac{10.0\ \cancel{mL}}{100\ \cancel{mL}} = 2.5\%$ (m/v) KOH solution (2 SFs)

    d. $C_2 = C_1 \times \dfrac{V_1}{V_2} = 15\% \times \dfrac{50.0\ \cancel{mL}}{250\ \cancel{mL}} = 3.0\%$ (m/v) H_2SO_4 solution (2 SFs)

7.57 $C_1V_1 = C_2V_2$ or $M_1V_1 = M_2V_2$

    a. $V_2 = V_1 \times \dfrac{M_1}{M_2} = 20.0\ mL \times \dfrac{6.0\ \cancel{M}}{1.5\ \cancel{M}} = 80.\ mL$ of diluted HCl solution (2 SFs)

    b. $V_2 = V_1 \times \dfrac{C_1}{C_2} = 50.0\ mL \times \dfrac{10.0\ \cancel{\%}}{2.0\ \cancel{\%}} = 250\ mL\ (2.5 \times 10^2\ mL)$ of diluted LiCl solution (2 SFs)

    c. $V_2 = V_1 \times \dfrac{M_1}{M_2} = 50.0\ mL \times \dfrac{6.00\ \cancel{M}}{0.500\ \cancel{M}} = 600.\ mL$ of diluted H_3PO_4 solution (3 SFs)

    d. $V_2 = V_1 \times \dfrac{C_1}{C_2} = 75\ mL \times \dfrac{12\ \cancel{\%}}{5.0\ \cancel{\%}} = 180\ mL\ (1.8 \times 10^2\ mL)$ of diluted glucose solution (2 SFs)

7.59 $M_1V_1 = M_2V_2$

    a. $V_1 = V_2 \times \dfrac{M_2}{M_1} = 255\ mL \times \dfrac{0.200\ \cancel{M}}{4.00\ \cancel{M}} = 12.8\ mL$ of the HNO_3 solution (3 SFs)

    b. $V_1 = V_2 \times \dfrac{M_2}{M_1} = 715\ mL \times \dfrac{0.100\ \cancel{M}}{6.00\ \cancel{M}} = 11.9\ mL$ of the $MgCl_2$ solution (3 SFs)

    c. $V_2 = 0.100\ \cancel{L} \times \dfrac{1000\ mL}{1\ \cancel{L}} = 100.\ mL$

        $V_1 = V_2 \times \dfrac{M_2}{M_1} = 100.\ mL \times \dfrac{0.150\ \cancel{M}}{8.00\ \cancel{M}} = 1.88\ mL$ of the KCl solution (3 SFs)

7.61   **a.** A solution cannot be separated by a semipermeable membrane.

      b. A suspension settles out upon standing.

7.63   **a.** When 1.0 mole of glycerol (a nonelectrolyte) dissolves in water, it does not dissociate into ions, and so will only produce 1.0 mole of particles. Similarly, 2.0 moles of ethylene glycol (also a nonelectrolyte) will produce 2.0 moles of particles in water. The ethylene glycol solution has more particles in 1.0 L of water, and thus will have a lower freezing point.

      b. When 0.50 mole of the strong electrolyte KCl dissolves in water, it will produce 1.0 mole of particles because each formula unit of KCl dissociates to give two particles, K^+ and Cl^-. When 0.50 mole of the strong electrolyte $MgCl_2$ dissolves in water, it will produce 1.5 moles of particles because each formula unit of $MgCl_2$ dissociates to give three particles, Mg^{2+} and $2Cl^-$. Thus, a solution of 0.50 mole of $MgCl_2$ in 2.0 L of water will have the lower freezing point.

7.65   **a.** The 10% (m/v) starch solution has the higher solute concentration, more solute particles, and the greater osmotic pressure.
      b. Initially, water will flow out of the 1% (m/v) starch solution into the more concentrated 10% (m/v) starch solution.
      c. The volume of the 10% (m/v) starch solution will increase due to inflow of water.

7.67   Water will flow from a region of higher solvent concentration (which corresponds to a lower solute concentration) to a region of lower solvent concentration (which corresponds to a higher solute concentration).
      a. The volume of compartment B will rise as water flows into compartment B, which contains the 10% (m/v) sucrose solution.
      b. The volume of compartment A will rise as water flows into compartment A, which contains the 8% (m/v) albumin solution.
      c. The volume of compartment B will rise as water flows into compartment B, which contains the 10% (m/v) starch solution.

7.69   A red blood cell has the same osmotic pressure as a 5% (m/v) glucose solution or a 0.9% (m/v) NaCl solution. In a hypotonic solution (lower osmotic pressure), solvent flows from the hypotonic solution into the red blood cell. When a red blood cell is placed into a hypertonic solution (higher osmotic pressure), solvent (water) flows from the red blood cell to the hypertonic solution. Isotonic solutions have the same osmotic pressure, and a red blood cell in an isotonic solution will not change volume because the flow of solvent into and out of the cell is equal.
      a. Distilled water is a hypotonic solution when compared with a red blood cell's contents.
      b. A 1% (m/v) glucose solution is a hypotonic solution.
      c. A 0.9% (m/v) NaCl solution is isotonic with a red blood cell's contents.
      d. A 15% (m/v) glucose solution is a hypertonic solution.

7.71   Colloids cannot pass through the semipermeable dialysis membrane; water and solutions freely pass through semipermeable membranes.
      a. Sodium and chloride ions will both pass through the membrane into the distilled water.
      b. The amino acid alanine can pass through a dialysis membrane; the colloid starch will not.
      c. Sodium and chloride ions will both be present in the water surrounding the dialysis bag; the colloid starch will not.
      d. Urea will diffuse through the dialysis bag into the surrounding water.

7.73   **a.** 1 (a solution will form because both the solute and the solvent are polar)
      b. 2 (two layers will form because one component is nonpolar and the other is polar)
      c. 1 (a solution will form because both the solute and the solvent are nonpolar)

7.75   **a.** 3 (a nonelectrolyte will show no dissociation)
      b. 1 (a weak electrolyte will show some dissociation, producing a few ions, but mostly remaining as molecules)
      c. 2 (a strong electrolyte will be completely dissociated into ions)

7.77   A "brine" saltwater solution has a high concentration of Na^+ and Cl^- ions, which is hypertonic to the cucumber. The skin of the cucumber acts like a semipermeable membrane; therefore, water flows from the more dilute solution inside the cucumber into the more concentrated brine solution that surrounds it. The loss of water causes the cucumber to become a wrinkled pickle.

7.79   **a.** 2; water will flow into the B (8% starch solution) side.
      b. 1; water will continue to flow equally in both directions; no change in volumes.
      c. 3; water will flow into the A (5% sucrose solution) side.
      d. 2; water will flow into the B (1% sucrose solution) side.

7.81 Because iodine is a nonpolar molecule, it will dissolve in hexane, a nonpolar solvent. Iodine does not dissolve in water because water is a polar solvent.

7.83 At 20 °C, KNO_3 has a solubility of 32 g of KNO_3 in 100 g of H_2O.

a. 200. g of H_2O will dissolve:

$$200. \text{ g } H_2O \times \frac{32 \text{ g } KNO_3}{100 \text{ g } H_2O} = 64 \text{ g of } KNO_3 \text{ (2 SFs)}$$

Because 2 g of KNO_3 is less than the maximum amount that can dissolve in 200. g of H_2O at 20 °C, the KNO_3 solution is unsaturated.

b. 50. g of H_2O will dissolve:

$$50. \text{ g } H_2O \times \frac{32 \text{ g } KNO_3}{100 \text{ g } H_2O} = 16 \text{ g of } KNO_3 \text{ (2 SFs)}$$

Because 19 g of KNO_3 exceeds the maximum amount that can dissolve in 50. g of H_2O at 20 °C, the KNO_3 solution is saturated, and excess undissolved KNO_3 will be present on the bottom of the container.

c. 150. g of H_2O will dissolve:

$$150. \text{ g } H_2O \times \frac{32 \text{ g } KNO_3}{100 \text{ g } H_2O} = 48 \text{ g of } KNO_3 \text{ (2 SFs)}$$

Because 68 g of KNO_3 exceeds the maximum amount that can dissolve in 150. g of H_2O at 20 °C, the KNO_3 solution is saturated, and excess undissolved KNO_3 will be present on the bottom of the container.

7.85 **a.** K^+ salts are soluble.

b. The SO_4^{2-} salt of Mg^{2+} is soluble.

c. The salt containing S^{2-} and Cu^{2+} is insoluble.

d. Salts containing NO_3^- ions are soluble.

e. The salt containing OH^- and Ca^{2+} is insoluble.

7.87 **a.** mass of solution $= 15.5$ g of $Na_2SO_4 + 75.5$ g of $H_2O = 91.0$ g of solution

$$\frac{15.5 \text{ g } Na_2SO_4}{91.0 \text{ g solution}} \times 100\% = 17.0\% \text{ (m/m) } Na_2SO_4 \text{ solution (3 SFs)}$$

7.89 **a.** $24 \text{ h} \times \dfrac{500 \text{ mL solution}}{12 \text{ h}} \times \dfrac{5.0 \text{ g amino acids}}{100 \text{ mL solution}} = 50 \text{ g of amino acids (1 SF)}$

$24 \text{ h} \times \dfrac{500 \text{ mL solution}}{12 \text{ h}} \times \dfrac{20 \text{ g glucose}}{100 \text{ mL solution}} = 200 \text{ g of glucose (1 SF)}$

$24 \text{ h} \times \dfrac{500 \text{ mL solution}}{12 \text{ h}} \times \dfrac{10 \text{ g lipid}}{100 \text{ mL solution}} = 100 \text{ g of lipid (1 SF)}$

b. $50 \text{ g amino acids (protein)} \times \dfrac{4 \text{ kcal}}{1 \text{ g protein}} = 200 \text{ kcal (1 SF)}$

$200 \text{ g glucose (carbohydrate)} \times \dfrac{4 \text{ kcal}}{1 \text{ g carbohydrate}} = 800 \text{ kcal (1 SF)}$

$100 \text{ g lipid (fat)} \times \dfrac{9 \text{ kcal}}{1 \text{ g fat}} = 900 \text{ kcal (1 SF)}$

For a total of $200 \text{ kcal} + 800 \text{ kcal} + 900 \text{ kcal} = 1900 \text{ kcal per day (2 SFs)}$

7.91 $4.5 \text{ mL propyl alcohol} \times \dfrac{100 \text{ mL solution}}{12 \text{ mL propyl alcohol}} = 38 \text{ mL of propyl alcohol solution (2 SFs)}$

7.93 $0.250 \text{ L solution} \times \dfrac{2.00 \text{ moles KCl}}{1 \text{ L solution}} \times \dfrac{74.6 \text{ g KCl}}{1 \text{ mole KCl}} = 37.3 \text{ g of KCl (3 SFs)}$

To make a 2.00 M KCl solution, weigh out 37.3 g of KCl (0.500 mole) and place in a volumetric flask. Add enough water to dissolve the KCl and give a final volume of 0.250 L.

7.95 mass of solution: 70.0 g of HNO_3 + 130.0 g of H_2O = 200.0 g of solution

a. $\dfrac{70.0 \text{ g } HNO_3}{200.0 \text{ g solution}} \times 100\% = 35.0\% \text{ (m/m) } HNO_3 \text{ solution (3 SFs)}$

b. $200.0 \text{ g solution} \times \dfrac{1 \text{ mL solution}}{1.21 \text{ g solution}} = 165 \text{ mL of solution (3 SFs)}$

c. $\dfrac{70.0 \text{ g } HNO_3}{165 \text{ mL solution}} \times 100\% = 42.4\% \text{ (m/v) } HNO_3 \text{ solution (3 SFs)}$

d. $\dfrac{70.0 \text{ g } HNO_3}{165 \text{ mL solution}} \times \dfrac{1 \text{ mole } HNO_3}{63.0 \text{ g } HNO_3} \times \dfrac{1000 \text{ mL solution}}{1 \text{ L solution}} = 6.73 \text{ M } HNO_3 \text{ solution (3 SFs)}$

7.97 **a.** $2.5 \text{ L solution} \times \dfrac{3.0 \text{ moles Al(NO}_3)_3}{1 \text{ L solution}} \times \dfrac{213.0 \text{ g Al(NO}_3)_3}{1 \text{ mole Al(NO}_3)_3}$

$= 1600 \text{ g of Al(NO}_3)_3 \text{ (2 SFs)}$

b. $75 \text{ mL solution} \times \dfrac{1 \text{ L solution}}{1000 \text{ mL solution}} \times \dfrac{0.50 \text{ mole } C_6H_{12}O_6}{1 \text{ L solution}} \times \dfrac{180.1 \text{ g } C_6H_{12}O_6}{1 \text{ mole } C_6H_{12}O_6}$

$= 6.8 \text{ g of } C_6H_{12}O_6 \text{ (2 SFs)}$

c. $235 \text{ mL solution} \times \dfrac{1 \text{ L solution}}{1000 \text{ mL solution}} \times \dfrac{1.80 \text{ moles LiCl}}{1 \text{ L solution}} \times \dfrac{42.4 \text{ g LiCl}}{1 \text{ mole LiCl}}$

$= 17.9 \text{ g of LiCl (3 SFs)}$

7.99 $M_1 V_1 = M_2 V_2$

a. $M_2 = M_1 \times \dfrac{V_1}{V_2} = 0.200 \text{ M} \times \dfrac{25.0 \text{ mL}}{50.0 \text{ mL}} = 0.100 \text{ M NaBr solution (3 SFs)}$

b. $M_2 = M_1 \times \dfrac{V_1}{V_2} = 1.20 \text{ M} \times \dfrac{15.0 \text{ mL}}{40.0 \text{ mL}} = 0.450 \text{ M } K_2SO_4 \text{ solution (3 SFs)}$

c. $M_2 = M_1 \times \dfrac{V_1}{V_2} = 6.00 \text{ M} \times \dfrac{75.0 \text{ mL}}{255 \text{ mL}} = 1.76 \text{ M NaOH solution (3 SFs)}$

7.101 $M_1V_1 = M_2V_2$

a. $V_2 = V_1 \times \dfrac{M_1}{M_2} = 25.0 \text{ mL} \times \dfrac{5.00 \text{ M}}{2.50 \text{ M}} = 50.0 \text{ mL of diluted HCl solution (3 SFs)}$

b. $V_2 = V_1 \times \dfrac{M_1}{M_2} = 25.0 \text{ mL} \times \dfrac{5.00 \text{ M}}{1.00 \text{ M}} = 125 \text{ mL of diluted HCl solution (3 SFs)}$

c. $V_2 = V_1 \times \dfrac{M_1}{M_2} = 25.0 \text{ mL} \times \dfrac{5.00 \text{ M}}{0.500 \text{ M}} = 250. \text{ mL of diluted HCl solution (3 SFs)}$

7.103 A solution with a high salt (solute) concentration is hypertonic to the cells of the flowers. Water (solvent) will flow out of the cells of the flowers into the hypertonic salt solution that surrounds them, resulting in "dried" flowers.

7.105 Drinking seawater, which has a high salt concentration and is hypertonic to body fluids, will cause water to flow out of the cells of the body, and further dehydrate a person.

7.107 **a.** mass of NaCl: $25.50 \text{ g} - 24.10 \text{ g} = 1.40 \text{ g of NaCl}$

mass of solution: $36.15 \text{ g} - 24.10 \text{ g} = 12.05 \text{ g of solution}$

mass percent (m/m): $\dfrac{1.40 \text{ g NaCl}}{12.05 \text{ g solution}} \times 100\% = 11.6\% \text{ (m/m) NaCl solution (3 SFs)}$

b. molarity (M): $\dfrac{1.40 \text{ g NaCl}}{10.0 \text{ mL solution}} \times \dfrac{1 \text{ mole NaCl}}{58.5 \text{ g NaCl}} \times \dfrac{1000 \text{ mL solution}}{1 \text{ L solution}}$

$= 2.39 \text{ M NaCl solution (3 SFs)}$

c. $M_1V_1 = M_2V_2 \quad M_2 = M_1 \times \dfrac{V_1}{V_2} = 2.39 \text{ M} \times \dfrac{10.0 \text{ mL}}{60.0 \text{ mL}} = 0.398 \text{ M NaCl solution (3 SFs)}$

7.109 At 18 °C, KF has a solubility of 92 g of KF in 100 g of H_2O.

a. 25 g of H_2O will dissolve:

$$25 \text{ g } H_2O \times \dfrac{92 \text{ g KF}}{100 \text{ g } H_2O} = 23 \text{ g of KF (2 SFs)}$$

Because 35 g of KF exceeds the maximum amount that can dissolve in 25 g of H_2O at 18 °C, the KF solution is saturated, and excess undissolved KF will be present on the bottom of the container.

b. 50. g of H_2O will dissolve:

$$50. \text{ g } H_2O \times \dfrac{92 \text{ g KF}}{100 \text{ g } H_2O} = 46 \text{ g of KF (2 SFs)}$$

Because 42 g of KF is less than the maximum amount that can dissolve in 50. g of H_2O at 18 °C, the solution is unsaturated.

   c. 150. g of H_2O will dissolve:

$$150. \text{ g } H_2O \times \frac{92 \text{ g KF}}{100 \text{ g } H_2O} = 140 \text{ g of KF (2 SFs)}$$

Because 145 g of KF exceeds the maximum amount that can dissolve in 150. g of H_2O at 18 °C, the KF solution is saturated, and excess undissolved KF will be present on the bottom of the container.

7.111 $15.2 \text{ g LiCl} \times \dfrac{1 \text{ mole LiCl}}{42.4 \text{ g LiCl}} \times \dfrac{1 \text{ L solution}}{1.75 \text{ moles LiCl}} \times \dfrac{1000 \text{ mL solution}}{1 \text{ L solution}}$

    $= 205 \text{ mL of LiCl solution (3 SFs)}$

7.113   **a.** A 2.0 M solution of KF (a strong electrolyte) will contain 4.0 moles of particles (K^+ and F^- ions) per liter of solution. A 1.0 M solution of $CaCl_2$ (also a strong electrolyte) will contain 3.0 moles of particles (Ca^{2+} and Cl^- ions) per liter of solution. Thus, the 2.0 M KF solution contains more particles, and so will have a lower freezing point.

  b. A 0.50 M solution of glucose (a nonelectrolyte) will contain 0.50 mole of particles (glucose molecules) per liter of solution. A 0.25 M solution of $CaCl_2$ (a strong electrolyte) will contain 0.75 mole of particles per liter of solution, since each $CaCl_2$ dissociates to form one Ca^{2+} and two Cl^- ions. Thus, the 0.25 M $CaCl_2$ solution contains more particles, and so will have a lower freezing point.

8

Acids and Bases

8.1   **a.**   Acids have a sour taste.
       b.   Acids neutralize bases.
       c.   Acids produce H^+ ions in water.
       d.   Potassium hydroxide is the name of a base.

8.3   Acids containing a simple nonmetal anion use the prefix *hydro*, followed by the name of the anion with its *ide* ending changed to *ic acid*. When the anion is an oxygen-containing polyatomic ion, the *ate* ending of the polyatomic anion is replaced with *ic acid*. Acids with one oxygen less than the common *ic acid* name are named as *ous acids*. Bases are named as ionic compounds containing hydroxide anions.
       a.   hydrochloric acid
       b.   calcium hydroxide
       c.   carbonic acid
       d.   nitric acid
       e.   sulfurous acid

8.5   **a.** $Mg(OH)_2$
       b.   HF
       c. H_3PO_4
       d.   LiOH
       e. NH_4OH
       f. H_2SO_4

8.7   A Brønsted–Lowry acid donates a proton (H^+), whereas a Brønsted–Lowry base accepts a proton.
       a.   HI is the acid (proton donor); H_2O is the base (proton acceptor).

       b. H_2O is the acid (proton donor); F^- is the base (proton acceptor).

8.9   To form the conjugate base, remove a proton (H^+) from the acid.

       a. F^-

       b. OH^-

       c. HCO_3^-

       d. SO_4^{2-}

8.11   To form the conjugate acid, add a proton (H^+) to the base.

       a. HCO_3^-

       b. H_3O^+

       c. H_3PO_4

       d.   HBr

8.13 The conjugate acid is a proton donor, and the conjugate base is a proton acceptor.

    a. In the reaction, the acid H_2CO_3 donates a proton to the base H_2O. The conjugate acid–base pairs are H_2CO_3/HCO_3^- and H_3O^+/H_2O.

    b. In the reaction, the acid NH_4^+ donates a proton to the base H_2O. The conjugate acid–base pairs are NH_4^+/NH_3 and H_3O^+/H_2O.

    c. In the reaction, the acid HCN donates a proton to the base NO_2^-. The conjugate acid–base pairs are HCN/CN^- and HNO_2/NO_2^-.

8.15 Use Table 8.3 to answer (the stronger acid will be closer to the top of the table).

    a. HBr is the stronger acid.

    b. HSO_4^- is the stronger acid.

    c. H_2CO_3 is the stronger acid.

8.17 Use Table 8.3 to answer (the weaker acid will be closer to the bottom of the table).

    a. HSO_4^- is the weaker acid.

    b. HNO_2 is the weaker acid.

    c. HCO_3^- is the weaker acid.

8.19 In pure water, a small fraction of the water molecules break apart to form H^+ and OH^-. The H^+ combines with H_2O to form H_3O^+. Every time a H^+ is formed, an OH^- is also formed. Therefore, the concentration of the two must be equal in pure water.

8.21 In an acidic solution, the $[H_3O^+]$ is greater than the $[OH^-]$, which means that the $[H_3O^+]$ is greater than 1×10^{-7} M and the $[OH^-]$ is less than 1×10^{-7} M.

8.23 The value of $K_w = [H_3O^+][OH^-] = 1.0\times10^{-14}$ at 25 °C.

If $[H_3O^+]$ needs to be calculated from $[OH^-]$, then rearranging the K_w for $[H_3O^+]$ gives

$$[H_3O^+] = \frac{1.0\times10^{-14}}{[OH^-]}.$$

If $[OH^-]$ needs to be calculated from $[H_3O^+]$, then rearranging the K_w for $[OH^-]$ gives

$$[OH^-] = \frac{1.0\times10^{-14}}{[H_3O^+]}.$$

A neutral solution has $[OH^-] = [H_3O^+] = 1\times10^{-7}$ M. If the $[OH^-] > [H_3O^+]$, the solution is basic; if the $[H_3O^+] > [OH^-]$, the solution is acidic.

    a. $[OH^-] = \dfrac{1.0\times10^{-14}}{[H_3O^+]} = \dfrac{1.0\times10^{-14}}{[2.0\times10^{-5}]} = 5.0\times10^{-10}$ M;

    since $[H_3O^+] > [OH^-]$, the solution is acidic.

b. $[OH^-] = \dfrac{1.0 \times 10^{-14}}{[H_3O^+]} = \dfrac{1.0 \times 10^{-14}}{[1.4 \times 10^{-9}]} = 7.1 \times 10^{-6}$ M;

since $[OH^-] > [H_3O^+]$, the solution is basic.

c. $[H_3O^+] = \dfrac{1.0 \times 10^{-14}}{[OH^-]} = \dfrac{1.0 \times 10^{-14}}{[8.0 \times 10^{-3}]} = 1.3 \times 10^{-12}$ M;

since $[OH^-] > [H_3O^+]$, the solution is basic.

d. $[H_3O^+] = \dfrac{1.0 \times 10^{-14}}{[OH^-]} = \dfrac{1.0 \times 10^{-14}}{[3.5 \times 10^{-10}]} = 2.9 \times 10^{-5}$ M;

since $[H_3O^+] > [OH^-]$, the solution is acidic.

8.25 The value of $K_w = [H_3O^+][OH^-] = 1.0 \times 10^{-14}$ at 25 °C.

When $[H_3O^+]$ is known, the $[OH^-]$ can be calculated by rearranging the K_w for $[OH^-]$:

$$[OH^-] = \dfrac{1.0 \times 10^{-14}}{[H_3O^+]}$$

a. $[OH^-] = \dfrac{1.0 \times 10^{-14}}{[H_3O^+]} = \dfrac{1.0 \times 10^{-14}}{[1.0 \times 10^{-5}]} = 1.0 \times 10^{-9}$ M (2 SFs)

b. $[OH^-] = \dfrac{1.0 \times 10^{-14}}{[H_3O^+]} = \dfrac{1.0 \times 10^{-14}}{[1.0 \times 10^{-8}]} = 1.0 \times 10^{-6}$ M (2 SFs)

c. $[OH^-] = \dfrac{1.0 \times 10^{-14}}{[H_3O^+]} = \dfrac{1.0 \times 10^{-14}}{[5.0 \times 10^{-10}]} = 2.0 \times 10^{-5}$ M (2 SFs)

d. $[OH^-] = \dfrac{1.0 \times 10^{-14}}{[H_3O^+]} = \dfrac{1.0 \times 10^{-14}}{[2.5 \times 10^{-2}]} = 4.0 \times 10^{-13}$ M (2 SFs)

8.27 The value of $K_w = [H_3O^+][OH^-] = 1.0 \times 10^{-14}$ at 25 °C.

When $[OH^-]$ is known, the $[H_3O^+]$ can be calculated by rearranging the K_w for $[H_3O^+]$:

$$[H_3O^+] = \dfrac{1.0 \times 10^{-14}}{[OH^-]}$$

a. $[H_3O^+] = \dfrac{1.0 \times 10^{-14}}{[OH^-]} = \dfrac{1.0 \times 10^{-14}}{[1.0 \times 10^{-11}]} = 1.0 \times 10^{-3}$ M (2 SFs)

b. $[H_3O^+] = \dfrac{1.0 \times 10^{-14}}{[OH^-]} = \dfrac{1.0 \times 10^{-14}}{[2.0 \times 10^{-9}]} = 5.0 \times 10^{-6}$ M (2 SFs)

c. $[H_3O^+] = \dfrac{1.0 \times 10^{-14}}{[OH^-]} = \dfrac{1.0 \times 10^{-14}}{[5.6 \times 10^{-3}]} = 1.8 \times 10^{-12}$ M (2 SFs)

d. $[H_3O^+] = \dfrac{1.0 \times 10^{-14}}{[OH^-]} = \dfrac{1.0 \times 10^{-14}}{[2.5 \times 10^{-2}]} = 4.0 \times 10^{-13}$ M (2 SFs)

8.29 In a neutral solution, the $[H_3O^+] = 1 \times 10^{-7}$ M.

$pH = -\log[H_3O^+] = -\log[1 \times 10^{-7}] = 7.0$. The pH value contains one *decimal place*, which represents the one significant figure in the coefficient 1.

8.31 An acidic solution has a pH less than 7.0. A basic solution has a pH greater than 7.0. A neutral solution has a pH equal to 7.0.
   a. basic (pH 7.38 > 7.0)
   b. acidic (pH 2.8 < 7.0)
   c. basic (pH 11.2 > 7.0)
   d. acidic (pH 5.54 < 7.0)
   e. acidic (pH 4.2 < 7.0)
   f. basic (pH 7.6 > 7.0)

8.33 $pH = -\log[H_3O^+]$

Since the value of $K_w = [H_3O^+][OH^-] = 1.0 \times 10^{-14}$ at 25 °C, if $[H_3O^+]$ needs to be calculated from $[OH^-]$, rearranging the K_w for $[H_3O^+]$ gives $[H_3O^+] = \dfrac{1.0 \times 10^{-14}}{[OH^-]}$.

   a. $pH = -\log[H_3O^+] = -\log[1 \times 10^{-4}] = 4.0$ (1 SF on the right of the decimal point)

   b. $pH = -\log[H_3O^+] = -\log[3 \times 10^{-9}] = 8.5$ (1 SF on the right of the decimal point)

   c. $[H_3O^+] = \dfrac{1.0 \times 10^{-14}}{[1 \times 10^{-5}]} = 1 \times 10^{-9}$ M

      $pH = -\log[1 \times 10^{-9}] = 9.0$ (1 SF on the right of the decimal point)

   d. $[H_3O^+] = \dfrac{1.0 \times 10^{-14}}{[2.5 \times 10^{-11}]} = 4.0 \times 10^{-4}$ M

      $pH = -\log[4.0 \times 10^{-4}] = 3.40$ (2 SFs on the right of the decimal point)

   e. $pH = -\log[H_3O^+] = -\log[6.7 \times 10^{-8}] = 7.17$ (2 SFs on the right of the decimal point)

   f. $[H_3O^+] = \dfrac{1.0 \times 10^{-14}}{[8.2 \times 10^{-4}]} = 1.2 \times 10^{-11}$ M

      $pH = -\log[1.2 \times 10^{-11}] = 10.92$ (2 SFs on the right of the decimal point)

8.35 On a calculator, pH is calculated by entering $-log$, followed by the coefficient *EE (EXP)* key and the power of 10 followed by the change sign $(+/-)$ key. On some calculators, the concentration is entered first (coefficient *EXP*–power) followed by *log* and $+/-$ key.

$$[H_3O^+] = \frac{1.0 \times 10^{-14}}{[OH^-]}; \ [OH^-] = \frac{1.0 \times 10^{-14}}{[H_3O^+]}; \ pH = -\log[H_3O^+]$$

$[H_3O^+]$	$[OH^-]$	pH	Acidic, Basic, or Neutral?
1×10^{-8} M	1×10^{-6} M	8.0	Basic
1×10^{-3} M	1×10^{-11} M	3.0	Acidic
2×10^{-5} M	5×10^{-10} M	4.7	Acidic
1×10^{-12} M	1×10^{-2} M	12.0	Basic
2.4×10^{-5} M	4.2×10^{-10} M	4.62	Acidic

8.37 Acids react with active metals to form $H_2(g)$ and a salt of the metal. The reaction of acids with carbonates yields $CO_2(g)$, H_2O, and a salt. In a neutralization reaction, an acid and a base react to form a salt and H_2O.

    a. $ZnCO_3(s) + 2HBr(aq) \rightarrow CO_2(g) + H_2O(l) + ZnBr_2(aq)$

    b. $Zn(s) + 2HCl(aq) \rightarrow H_2(g) + ZnCl_2(aq)$

    c. $HCl(aq) + NaHCO_3(s) \rightarrow CO_2(g) + H_2O(l) + NaCl(aq)$

    d. $H_2SO_4(aq) + Mg(OH)_2(s) \rightarrow MgSO_4(aq) + 2H_2O(l)$

8.39 In balancing a neutralization equation, the number of H^+ and OH^- must be equalized by placing coefficients in front of the formulas for the acid and base.

    a. $2HCl(aq) + Mg(OH)_2(s) \rightarrow MgCl_2(aq) + 2H_2O(l)$

    b. $H_3PO_4(aq) + 3LiOH(aq) \rightarrow Li_3PO_4(aq) + 3H_2O(l)$

8.41 The products of a neutralization are water and a salt. In balancing a neutralization equation, the number of H^+ and OH^- must be equalized by placing coefficients in front of the formulas for the acid and base.

    a. $H_2SO_4(aq) + 2NaOH(aq) \rightarrow Na_2SO_4(aq) + 2H_2O(l)$

    b. $3HCl(aq) + Fe(OH)_3(s) \rightarrow FeCl_3(aq) + 3H_2O(l)$

    c. $H_2CO_3(aq) + Mg(OH)_2(s) \rightarrow MgCO_3(s) + 2H_2O(l)$

8.43 In the titration equation, one mole of HCl reacts with one mole of NaOH.

$$28.6 \text{ mL NaOH solution} \times \frac{1 \text{ L solution}}{1000 \text{ mL solution}} \times \frac{0.145 \text{ mole NaOH}}{1 \text{ L solution}} \times \frac{1 \text{ mole HCl}}{1 \text{ mole NaOH}}$$

$$= 0.004\ 15 \text{ mole of HCl}$$

$$5.00 \text{ mL HCl solution} \times \frac{1 \text{ L solution}}{1000 \text{ mL solution}} = 0.005\ 00 \text{ L of HCl solution}$$

$$\text{molarity (M) of HCl} = \frac{\text{moles of solute}}{\text{liters of solution}} = \frac{0.004\ 15 \text{ mole HCl}}{0.005\ 00 \text{ L solution}}$$

$$= 0.830 \text{ M HCl solution (3 SFs)}$$

8.45 In the titration equation, one mole of H_2SO_4 reacts with two moles of KOH.

$$38.2 \; \cancel{mL \; KOH \; solution} \times \frac{1 \; \cancel{L \; solution}}{1000 \; \cancel{mL \; solution}} \times \frac{0.163 \; \cancel{mole \; KOH}}{1 \; \cancel{L \; solution}} \times \frac{1 \; mole \; H_2SO_4}{2 \; \cancel{moles \; KOH}}$$

$$= 0.003 \; 11 \; mole \; of \; H_2SO_4$$

$$25.0 \; \cancel{mL \; H_2SO_4 \; solution} \times \frac{1 \; L \; solution}{1000 \; \cancel{mL \; solution}} = 0.0250 \; L \; of \; H_2SO_4 \; solution$$

$$molarity \; (M) \; of \; H_2SO_4 = \frac{moles \; of \; solute}{liters \; of \; solution} = \frac{0.003 \; 11 \; mole \; H_2SO_4}{0.0250 \; L \; solution}$$

$$= 0.124 \; M \; H_2SO_4 \; solution \; (3 \; SFs)$$

8.47 A buffer system contains a weak acid and a salt containing its conjugate base (or a weak base and a salt containing its conjugate acid).
  a. This is not a buffer system because it only contains the strong base NaOH and the neutral salt NaCl.
  b. This is a buffer system; it contains the weak acid H_2CO_3 and a salt containing its conjugate base HCO_3^-.
  c. This is a buffer system; it contains the weak acid HF and a salt containing its conjugate base F^-.
  d. This is not a buffer system because it only contains the neutral salts KCl and NaCl.

8.49 **a.** A buffer system keeps the pH of a solution constant.
  b. A buffer needs a salt to supply more of the conjugate base of the weak acid to neutralize any H_3O^+ added.
  c. When H_3O^+ is added to the buffer system, the F^- from the salt NaF reacts with the acid to neutralize it.
$$F^-(aq) + H_3O^+(aq) \rightarrow HF(aq) + H_2O(l)$$
  d. When OH^- is added to the buffer system, the weak acid HF reacts with the OH^- to neutralize it.
$$HF(aq) + OH^-(aq) \rightarrow F^-(aq) + H_2O(l)$$

8.51 **a.** base; lithium hydroxide
  b. salt; calcium nitrate
  c. acid; hydrobromic acid
  d. base; barium hydroxide
  e. acid; carbonic acid

8.53

Acid	Conjugate Base
H_2O	OH^-
HCN	CN^-
HNO_2	NO_2^-
H_3PO_4	$H_2PO_4^-$

8.55   **a.**   This diagram represents a weak acid; only a few HX molecules dissociate into H_3O^+ and X^- ions.

      b.   This diagram represents a strong acid; all of the HX molecules dissociate into H_3O^+ and X^- ions.

      c.   This diagram represents a weak acid; only a few HX molecules dissociate into H_3O^+ and X^- ions.

8.57   **a.**   Hyperventilation will lower the CO_2 level in the blood, which lowers the $[H_2CO_3]$, which decreases the $[H_3O^+]$ and increases the blood pH.

      b.   Breathing into a paper bag will increase the CO_2 level in the blood, increase the $[H_2CO_3]$, increase $[H_3O^+]$ and lower the blood pH back toward the normal range.

8.59   **a.**   sulfuric acid
      b.   rubidium hydroxide
      c.   calcium hydroxide
      d.   hydroiodic acid

8.61   An acidic solution has a pH less than 7.0. A neutral solution has a pH equal to 7.0. A basic solution has a pH greater than 7.0.
      a.   acidic (pH $5.2 < 7.0$)
      b.   basic (pH $7.5 > 7.0$)
      c.   acidic (pH $3.8 < 7.0$)
      d.   acidic (pH $2.5 < 7.0$)
      e.   basic (pH $12.0 > 7.0$)

8.63   Both strong and weak acids produce H_3O^+ in water. They both neutralize bases, turn litmus red and phenolphthalein clear. Both taste sour and are electrolytes in solution. However, weak acids are only slightly dissociated, to give an aqueous solution of only a few ions, making them weak electrolytes. Strong acids are nearly completely dissociated, giving only ions in solution, which makes them strong electrolytes.

8.65   **a.** $Mg(OH)_2$ is considered a strong base because all the $Mg(OH)_2$ that dissolves is completely dissociated in aqueous solution.

      b. $Mg(OH)_2(s) + 2HCl(aq) \rightarrow MgCl_2(aq) + 2H_2O(l)$

8.67 $[H_3O^+] = \dfrac{1.0 \times 10^{-14}}{[OH^-]}$; $pH = -\log[H_3O^+]$

      a. $pH = -\log[H_3O^+] = -\log[1.0 \times 10^{-8}] = 8.00$ (2 SFs on the right of the decimal point)

      b. $pH = -\log[5.0 \times 10^{-2}] = 1.30$ (2 SFs on the right of the decimal point)

      c. $[H_3O^+] = \dfrac{1.0 \times 10^{-14}}{[3.5 \times 10^{-4}]} = 2.9 \times 10^{-11}$ M

         $pH = -\log[2.9 \times 10^{-11}] = 10.54$ (2 SFs on the right of the decimal point)

d. $[H_3O^+] = \dfrac{1.0 \times 10^{-14}}{[0.005]} = 2 \times 10^{-12}$ M

$pH = -\log[2 \times 10^{-12}] = 11.7$ (1 SF on the right of the decimal point)

8.69 $[H_3O^+] = \dfrac{1.0 \times 10^{-14}}{[OH^-]}$; $pH = -\log[H_3O^+]$

a. $[H_3O^+] = \dfrac{1.0 \times 10^{-14}}{[1.0 \times 10^{-7}]} = 1.0 \times 10^{-7}$ M

$pH = -\log[1.0 \times 10^{-7}] = 7.00$ (2 SFs on the right of the decimal point)

b. $pH = -\log[H_3O^+] = -\log[4.2 \times 10^{-3}] = 2.38$ (2 SFs on the right of the decimal point)

c. $pH = -\log[H_3O^+] = -\log[0.0001] = 4.0$ (1 SF on the right of the decimal point)

d. $[H_3O^+] = \dfrac{1.0 \times 10^{-14}}{[8.5 \times 10^{-9}]} = 1.2 \times 10^{-6}$ M

$pH = -\log[1.2 \times 10^{-6}] = 5.92$ (2 SFs on the right of the decimal point)

8.71 If the pH is given, the $[H_3O^+]$ can be found by using the relationship $[H_3O^+] = 1 \times 10^{-pH}$.

The $[OH^-]$ can be found by rearranging $K_w = [H_3O^+][OH^-] = 1 \times 10^{-14}$.

a. $pH = 3.0$; $[H_3O^+] = 1 \times 10^{-pH} = 1 \times 10^{-3.0} = 1 \times 10^{-3}$ M (1 SF)

$[OH^-] = \dfrac{1.0 \times 10^{-14}}{[H_3O^+]} = \dfrac{1.0 \times 10^{-14}}{[1 \times 10^{-3}]} = 1 \times 10^{-11}$ M (1 SF)

b. $pH = 6.00$; $[H_3O^+] = 1 \times 10^{-pH} = 1 \times 10^{-6.00} = 1.0 \times 10^{-6}$ M (2 SFs)

$[OH^-] = \dfrac{1.0 \times 10^{-14}}{[H_3O^+]} = \dfrac{1.0 \times 10^{-14}}{[1.0 \times 10^{-6}]} = 1.0 \times 10^{-8}$ M (2 SFs)

c. $pH = 8.0$; $[H_3O^+] = 1 \times 10^{-pH} = 1 \times 10^{-8.0} = 1 \times 10^{-8}$ M (1 SF)

$[OH^-] = \dfrac{1.0 \times 10^{-14}}{[H_3O^+]} = \dfrac{1.0 \times 10^{-14}}{[1 \times 10^{-8}]} = 1 \times 10^{-6}$ M (1 SF)

d. $pH = 11.0$; $[H_3O^+] = 1 \times 10^{-pH} = 1 \times 10^{-11.0} = 1 \times 10^{-11}$ M (1 SF)

$[OH^-] = \dfrac{1.0 \times 10^{-14}}{[H_3O^+]} = \dfrac{1.0 \times 10^{-14}}{[1 \times 10^{-11}]} = 1 \times 10^{-3}$ M (1 SF)

e. $pH = 9.20$; $[H_3O^+] = 1 \times 10^{-pH} = 1 \times 10^{-9.20} = 6.3 \times 10^{-10}$ M (2 SFs)

$[OH^-] = \dfrac{1.0 \times 10^{-14}}{[H_3O^+]} = \dfrac{1.0 \times 10^{-14}}{[6.3 \times 10^{-10}]} = 1.6 \times 10^{-5}$ M (2 SFs)

8.73 **a.** Solution A, with a pH of 4.0, is more acidic than solution B.

   b. In solution A, the $[H_3O^+] = 1 \times 10^{-pH} = 1 \times 10^{-4.0} = 1 \times 10^{-4}$ M (1 SF)

   In solution B, the $[H_3O^+] = 1 \times 10^{-pH} = 1 \times 10^{-6.0} = 1 \times 10^{-6}$ M (1 SF)

   c. In solution A, the $[OH^-] = \dfrac{1.0 \times 10^{-14}}{[H_3O^+]} = \dfrac{1.0 \times 10^{-14}}{[1 \times 10^{-4}]} = 1 \times 10^{-10}$ M (1 SF)

   In solution B, the $[OH^-] = \dfrac{1.0 \times 10^{-14}}{[H_3O^+]} = \dfrac{1.0 \times 10^{-14}}{[1 \times 10^{-6}]} = 1 \times 10^{-8}$ M (1 SF)

8.75 This buffer solution is made from the weak acid H_3PO_4 and a salt containing its conjugate base $H_2PO_4^-$.

   a. Acid added: $H_2PO_4^-(aq) + H_3O^+(aq) \rightarrow H_3PO_4(aq) + H_2O(l)$

   b. Base added: $H_3PO_4(aq) + OH^-(aq) \rightarrow H_2PO_4^-(aq) + H_2O(l)$

8.77 **a.** In the titration equation, one mole of HCl reacts with one mole of NaOH.

   $$HCl(aq) + NaOH(aq) \rightarrow NaCl(aq) + H_2O(l)$$

   $$25.0 \text{ mL HCl solution} \times \frac{1 \text{ L HCl solution}}{1000 \text{ mL HCl solution}} \times \frac{0.288 \text{ mole HCl}}{1 \text{ L HCl solution}} \times \frac{1 \text{ mole NaOH}}{1 \text{ mole HCl}}$$

   $$\times \frac{1 \text{ L NaOH solution}}{0.150 \text{ mole NaOH}} \times \frac{1000 \text{ mL NaOH solution}}{1 \text{ L NaOH solution}} = 48.0 \text{ mL of NaOH solution (3 SFs)}$$

   b. In the titration equation, one mole of H_2SO_4 reacts with two moles of NaOH.

   $$H_2SO_4(aq) + 2NaOH(aq) \rightarrow Na_2SO_4(aq) + 2H_2O(l)$$

   $$10.0 \text{ mL } H_2SO_4 \text{ solution} \times \frac{1 \text{ L } H_2SO_4 \text{ solution}}{1000 \text{ mL } H_2SO_4 \text{ solution}} \times \frac{0.560 \text{ mole } H_2SO_4}{1 \text{ L } H_2SO_4 \text{ solution}} \times \frac{2 \text{ moles NaOH}}{1 \text{ mole } H_2SO_4}$$

   $$\times \frac{1 \text{ L NaOH solution}}{0.150 \text{ mole NaOH}} \times \frac{1000 \text{ mL NaOH solution}}{1 \text{ L NaOH solution}} = 74.7 \text{ mL of NaOH solution (3 SFs)}$$

   c. In the titration equation, one mole of HBr reacts with one mole of NaOH.

   $$HBr(aq) + NaOH(aq) \rightarrow NaBr(aq) + H_2O(l)$$

   $$5.00 \text{ mL HBr solution} \times \frac{1 \text{ L HBr solution}}{1000 \text{ mL HBr solution}} \times \frac{0.618 \text{ mole HBr}}{1 \text{ L HBr solution}} \times \frac{1 \text{ mole NaOH}}{1 \text{ mole HBr}}$$

   $$\times \frac{1 \text{ L NaOH solution}}{0.150 \text{ mole NaOH}} \times \frac{1000 \text{ mL NaOH solution}}{1 \text{ L NaOH solution}} = 20.6 \text{ mL of NaOH solution (3 SFs)}$$

8.79 In the titration equation, one mole of H_2SO_4 reacts with two moles of NaOH.

$$45.6 \text{ mL NaOH solution} \times \frac{1 \text{ L solution}}{1000 \text{ mL solution}} \times \frac{0.205 \text{ mole NaOH}}{1 \text{ L solution}} \times \frac{1 \text{ mole } H_2SO_4}{2 \text{ moles NaOH}}$$

$$= 0.004 \ 67 \text{ mole of } H_2SO_4$$

$$20.0 \text{ mL } H_2SO_4 \text{ solution} \times \frac{1 \text{ L solution}}{1000 \text{ mL solution}} = 0.0200 \text{ L of } H_2SO_4 \text{ solution}$$

$$\text{molarity (M) of } H_2SO_4 = \frac{\text{moles of solute}}{\text{liters of solution}} = \frac{0.004 \ 67 \text{ mole } H_2SO_4}{0.0200 \text{ L solution}}$$

$$= 0.234 \text{ M } H_2SO_4 \text{ solution (3 SFs)}$$

8.81 **a.** To form the conjugate base, remove a proton (H^+) from the acid.

   1. HS^- **2.** $H_2PO_4^-$ **3.** CO_3^{2-}

   b. HCO_3^- (see Table 8.3, the weakest acid will be closest to the bottom of the table)

   c. H_3PO_4 (see Table 8.3, the strongest acid will be closest to the top of the table)

8.83 **a.** $ZnCO_3(s) + H_2SO_4(aq) \rightarrow CO_2(g) + H_2O(l) + ZnSO_4(aq)$

   b. $2Al(s) + 6HCl(aq) \rightarrow 3H_2(g) + 2AlCl_3(aq)$

   c. $2H_3PO_4(aq) + 3Ca(OH)_2(s) \rightarrow Ca_3(PO_4)_2(aq) + 6H_2O(l)$

   d. $KHCO_3(s) + HNO_3(aq) \rightarrow CO_2(g) + H_2O(l) + KNO_3(aq)$

8.85 $2.5 \text{ g HCl} \times \dfrac{1 \text{ mole HCl}}{36.5 \text{ g HCl}} = 0.069 \text{ mole of HCl}$

$$425 \text{ mL HCl solution} \times \frac{1 \text{ L solution}}{1000 \text{ mL solution}} = 0.425 \text{ L of HCl solution}$$

$$\text{molarity (M) of HCl} = \frac{\text{moles of solute}}{\text{liters of solution}} = \frac{0.069 \text{ mole HCl}}{0.425 \text{ L solution}} = 0.16 \text{ M HCl solution (2 SFs)}$$

Since HCl is a strong acid, the $[H_3O^+]$ is also 0.16 M.

$$pH = -\log[H_3O^+] = -\log[0.16] = 0.80 \text{ (2 SFs on the right of the decimal point)}$$

8.87 **a.** $H_3PO_4(aq) + 3NaOH(aq) \rightarrow Na_3PO_4(aq) + 3H_2O(l)$

   b. In the titration equation, one mole of H_3PO_4 reacts with three moles of NaOH.

$$16.4 \text{ mL NaOH solution} \times \frac{1 \text{ L NaOH solution}}{1000 \text{ mL NaOH solution}} \times \frac{0.204 \text{ mole NaOH}}{1 \text{ L NaOH solution}} \times \frac{1 \text{ mole } H_3PO_4}{3 \text{ moles NaOH}}$$

$$= 1.12 \times 10^{-3} \text{ mole of } H_3PO_4 \text{ (3 SFs)}$$

$$50.0 \ \text{mL} \ \cancel{H_3PO_4 \ \text{solution}} \times \frac{1 \ \text{L solution}}{1000 \ \cancel{\text{mL solution}}} = 0.0500 \ \text{L of} \ H_3PO_4 \ \text{solution}$$

$$\text{molarity (M) of} \ H_3PO_4 = \frac{\text{moles of solute}}{\text{liters of solution}} = \frac{1.12 \times 10^{-3} \ \text{mole} \ H_3PO_4}{0.0500 \ \text{L solution}}$$

$$= 0.0224 \ \text{M} \ H_3PO_4 \ \text{solution (3 SFs)}$$

8.89 **a.** The $[H_3O^+]$ can be found by using the relationship $[H_3O^+] = 1 \times 10^{-pH}$.

$$[H_3O^+] = 1 \times 10^{-pH} = 1 \times 10^{-4.2} = 6 \times 10^{-5} \ \text{M (1 SF)}$$

$$[OH^-] = \frac{1.0 \times 10^{-14}}{[H_3O^+]} = \frac{1.0 \times 10^{-14}}{[6 \times 10^{-5}]} = 2 \times 10^{-10} \ \text{M (1 SF)}$$

b. $[H_3O^+] = 1 \times 10^{-pH} = 1 \times 10^{-6.5} = 3 \times 10^{-7} \ \text{M (1 SF)}$

$$[OH^-] = \frac{1.0 \times 10^{-14}}{[H_3O^+]} = \frac{1.0 \times 10^{-14}}{[3 \times 10^{-7}]} = 3 \times 10^{-8} \ \text{M (1 SF)}$$

c. In the titration equation, one mole of $CaCO_3$ reacts with two moles of acid HA.

$$1.0 \ \cancel{\text{kL solution}} \times \frac{1000 \ \cancel{\text{L solution}}}{1 \ \cancel{\text{kL solution}}} \times \frac{6 \times 10^{-5} \ \cancel{\text{mole HA}}}{1 \ \cancel{\text{L solution}}} \times \frac{1 \ \text{mole} \ \cancel{CaCO_3}}{2 \ \cancel{\text{moles HA}}}$$

$$\times \frac{100.1 \ \text{g} \ CaCO_3}{1 \ \text{mole} \ \cancel{CaCO_3}} = 3 \ \text{g of} \ CaCO_3 \ \text{(1 SF)}$$

Answers to Combining Ideas from Chapters 6 to 8

CI.15 a. CH_4

$$\begin{matrix} & H \\ & \overset{\cdot\cdot}{} \\ H:&\overset{\cdot\cdot}{C}:H \\ & \overset{\cdot\cdot}{H} \end{matrix} = \begin{matrix} & H \\ & | \\ H-&C-H \\ & | \\ & H \end{matrix}$$

b. $7.0\times10^6 \text{ gal} \times \dfrac{4 \text{ qt}}{1 \text{ gal}} \times \dfrac{946 \text{ mL}}{1 \text{ qt}} \times \dfrac{0.45 \text{ g}}{1 \text{ mL}} \times \dfrac{1 \text{ kg}}{1000 \text{ g}} = 1.2\times10^7 \text{ kg of LNG (2 SFs)}$

c. $7.0\times10^6 \text{ gal} \times \dfrac{4 \text{ qt}}{1 \text{ gal}} \times \dfrac{946 \text{ mL}}{1 \text{ qt}} \times \dfrac{0.45 \text{ g}}{1 \text{ mL}} \times \dfrac{1 \text{ mole } CH_4}{16.0 \text{ g}} \times \dfrac{22.4 \text{ L } CH_4 \text{ (STP)}}{1 \text{ mole } CH_4}$

$= 1.7\times10^{10} \text{ L of LNG at STP (2 SFs)}$

d. $CH_4(g) + 2O_2(g) \xrightarrow{\Delta} CO_2(g) + 2H_2O(g) + 883 \text{ kJ}$

e. $7.0\times10^6 \text{ gal} \times \dfrac{4 \text{ qt}}{1 \text{ gal}} \times \dfrac{946 \text{ mL}}{1 \text{ qt}} \times \dfrac{0.45 \text{ g}}{1 \text{ mL}} \times \dfrac{1 \text{ mole } CH_4}{16.0 \text{ g}}$

$\times \dfrac{2 \text{ moles } O_2}{1 \text{ mole } CH_4} \times \dfrac{32.0 \text{ g } O_2}{1 \text{ mole } O_2} \times \dfrac{1 \text{ kg } O_2}{1000 \text{ g } O_2}$

$= 4.8\times10^7 \text{ kg of } O_2 \text{ (2 SFs)}$

f. $7.0\times10^6 \text{ gal} \times \dfrac{4 \text{ qt}}{1 \text{ gal}} \times \dfrac{946 \text{ mL}}{1 \text{ qt}} \times \dfrac{0.45 \text{ g}}{1 \text{ mL}} \times \dfrac{1 \text{ mole } CH_4}{16.0 \text{ g}} \times \dfrac{883 \text{ kJ}}{1 \text{ mole } CH_4}$

$= 6.6\times10^{11} \text{ kJ (2 SFs)}$

CI.17 a. Sodium hypochlorite (NaClO) is an ionic compound, composed of sodium (Na^+) and hypochlorite (ClO^-) ions.

b. volume of solution $= 1.42 \text{ gal bleach} \times \dfrac{4 \text{ qt solution}}{1 \text{ gal solution}} \times \dfrac{946 \text{ mL solution}}{1 \text{ qt solution}}$

$= 5.37\times10^3 \text{ mL of solution (3 SFs)}$

mass/volume $\% = \dfrac{\text{mass of solute (g)}}{\text{volume of solution (mL)}} \times 100\%$

$= \dfrac{282 \text{ g NaClO}}{5.37\times10^3 \text{ mL solution}} \times 100\% = 5.25\% \text{ (m/v) NaClO solution (3 SFs)}$

c. $2NaOH(aq) + Cl_2(g) \rightarrow NaClO(aq) + NaCl(aq) + H_2O(l)$

d. **Given** 1 bottle of bleach   **Need** liters of Cl_2 gas at STP

Plan bottle $\rightarrow$ gallons of bleach $\rightarrow$ grams of NaClO $\rightarrow$ moles of NaClO

$\rightarrow$ moles of Cl_2 $\rightarrow$ liters of Cl_2 gas (STP)

$$1 \text{ bottle bleach} \times \frac{1.42 \text{ gal bleach}}{1 \text{ bottle bleach}} \times \frac{282 \text{ g NaClO}}{1.42 \text{ gal bleach}} \times \frac{1 \text{ mole NaClO}}{74.5 \text{ g NaClO}}$$

$$\times \frac{1 \text{ mole Cl}_2}{1 \text{ mole NaClO}} \times \frac{22.4 \text{ L Cl}_2 \text{ (STP)}}{1 \text{ mole Cl}_2}$$

$$= 84.8 \text{ L of Cl}_2 \text{ at STP (3 SFs)}$$

e. If the pH is given, the $[H_3O^+]$ can be found by using the relationship $[H_3O^+] = 1 \times 10^{-pH}$.

The $[OH^-]$ can be found by rearranging $K_w = [H_3O^+][OH^-] = 1 \times 10^{-14}$.

$$pH = 10.3; \quad [H_3O^+] = 1 \times 10^{-pH} = 1 \times 10^{-10.3} = 5 \times 10^{-11} \text{ M (1 SF)}$$

$$[OH^-] = \frac{1.0 \times 10^{-14}}{[H_3O^+]} = \frac{1.0 \times 10^{-14}}{[5 \times 10^{-11}]} = 2 \times 10^{-4} \text{ M (1 SF)}$$

CI.19 **a.** $2M(s) + 6HCl(aq) \rightarrow 2MCl_3(aq) + 3H_2(g)$

b. $34.8 \text{ mL solution} \times \dfrac{1 \text{ L solution}}{1000 \text{ mL solution}} \times \dfrac{0.520 \text{ mole HCl}}{1 \text{ L solution}} \times \dfrac{3 \text{ moles H}_2}{6 \text{ moles HCl}}$

$$\times \frac{22.4 \text{ L H}_2 \text{(STP)}}{1 \text{ mole H}_2} \times \frac{1000 \text{ mL H}_2}{1 \text{ L H}_2}$$

$$= 203 \text{ mL of H}_2 \text{ at STP (3 SFs)}$$

c. $34.8 \text{ mL solution} \times \dfrac{1 \text{ L solution}}{1000 \text{ mL solution}} \times \dfrac{0.520 \text{ mole HCl}}{1 \text{ L solution}} \times \dfrac{2 \text{ moles M}}{6 \text{ moles HCl}}$

$$= 6.03 \times 10^{-3} \text{ mole of M (3 SFs)}$$

d. $\dfrac{0.420 \text{ g M}}{6.03 \times 10^{-3} \text{ mole M}} = 69.7 \text{ g/mole of M (3 SFs)}$

e. The metal is gallium, Ga.

f. $2Ga(s) + 6HCl(aq) \rightarrow 2GaCl_3(aq) + 3H_2(g)$

9
Nuclear Radiation

9.1   **a.** An alpha particle and a helium nucleus both contain two protons and two neutrons.
    b. α, $_2^4\text{He}$
    c. An alpha particle is emitted from an unstable nucleus during radioactive decay.

9.3   **a.** $_{19}^{39}\text{K}$, $_{19}^{40}\text{K}$, $_{19}^{41}\text{K}$
    b. Each isotope has 19 protons and 19 electrons, but they differ in the number of neutrons present. Potassium-39 has $39 - 19 = 20$ neutrons, potassium-40 has $40 - 19 = 21$ neutrons, and potassium-41 has $41 - 19 = 22$ neutrons.

9.5

Medical Use	Atomic Symbol	Mass Number	Number of Protons	Number of Neutrons
Heart imaging	$_{81}^{201}\text{Tl}$	201	81	120
Radiation therapy	$_{27}^{60}\text{Co}$	60	27	33
Abdominal scan	$_{31}^{67}\text{Ga}$	67	31	36
Hyperthyroidism	$_{53}^{131}\text{I}$	131	53	78
Leukemia treatment	$_{15}^{32}\text{P}$	32	15	17

9.7   **a.** α, $_2^4\text{He}$
    b. n, $_0^1n$
    c. β, $_{-1}^0e$
    d. $_7^{15}\text{N}$
    e. $_{53}^{125}\text{I}$

9.9   **a.** beta particle (β, $_{-1}^0e$)
    b. alpha particle (α, $_2^4\text{He}$)
    c. neutron (n, $_0^1n$)
    d. sodium-24 ($_{11}^{24}\text{Na}$)
    e. carbon-14 ($_6^{14}\text{C}$)

9.11   **a.** Because beta particles have less mass and move faster than alpha particles, beta radiation can penetrate farther into body tissue.
    b. Ionizing radiation breaks bonds and forms reactive species that cause undesirable reactions in the cells of the body.
    c. Radiation technicians leave the room to increase the distance between themselves and the radiation. A thick wall of concrete or lead acts to shield them further.
    d. Wearing gloves when handling radioisotopes shields the skin from α and β radiation.

9.13 The mass number of the radioactive atom is reduced by 4 when an alpha particle ($^4_2$He) is emitted. The unknown product will have an atomic number that is 2 less than the atomic number of the radioactive atom.

    a. $^{208}_{84}\text{Po} \rightarrow {}^{204}_{82}\text{Pb} + {}^4_2\text{He}$

    b. $^{232}_{90}\text{Th} \rightarrow {}^{228}_{88}\text{Ra} + {}^4_2\text{He}$

    c. $^{251}_{102}\text{No} \rightarrow {}^{247}_{100}\text{Fm} + {}^4_2\text{He}$

    d. $^{220}_{86}\text{Rn} \rightarrow {}^{216}_{84}\text{Po} + {}^4_2\text{He}$

9.15 The mass number of the radioactive atom is not changed when a beta particle ($^0_{-1}e$) is emitted. The unknown product will have an atomic number that is 1 higher than the atomic number of the radioactive atom.

    a. $^{25}_{11}\text{Na} \rightarrow {}^{25}_{12}\text{Mg} + {}^0_{-1}e$

    b. $^{20}_{8}\text{O} \rightarrow {}^{20}_{9}\text{F} + {}^0_{-1}e$

    c. $^{92}_{38}\text{Sr} \rightarrow {}^{92}_{39}\text{Y} + {}^0_{-1}e$

    d. $^{60}_{26}\text{Fe} \rightarrow {}^{60}_{27}\text{Co} + {}^0_{-1}e$

9.17 The mass number of the radioactive atom is not changed when a positron ($^0_{+1}e$) is emitted. The unknown product will have an atomic number that is 1 less than the atomic number of the radioactive atom.

    a. $^{26}_{14}\text{Si} \rightarrow {}^{26}_{13}\text{Al} + {}^0_{+1}e$

    b. $^{54}_{27}\text{Co} \rightarrow {}^{54}_{26}\text{Fe} + {}^0_{+1}e$

    c. $^{77}_{37}\text{Rb} \rightarrow {}^{77}_{36}\text{Kr} + {}^0_{+1}e$

    d. $^{93}_{45}\text{Rh} \rightarrow {}^{93}_{44}\text{Ru} + {}^0_{+1}e$

9.19 Balance the mass numbers and the atomic numbers in each nuclear equation.

    a. $^{28}_{13}\text{Al} \rightarrow {}^{28}_{14}\text{Si} + {}^0_{-1}e$ $? = {}^{28}_{14}\text{Si}$ beta decay

    b. $^{180m}_{73}\text{Ta} \rightarrow {}^{180}_{73}\text{Ta} + {}^0_0\gamma$ $? = {}^0_0\gamma$ gamma emission

    c. $^{66}_{29}\text{Cu} \rightarrow {}^{66}_{30}\text{Zn} + {}^0_{-1}e$ $? = {}^0_{-1}e$ beta decay

    d. $^{238}_{92}\text{U} \rightarrow {}^4_2\text{He} + {}^{234}_{90}\text{Th}$ $? = {}^{238}_{92}\text{U}$ alpha decay

    e. $^{188}_{80}\text{Hg} \rightarrow {}^{188}_{79}\text{Au} + {}^0_{+1}e$ $? = {}^{188}_{79}\text{Au}$ positron emission

9.21 Balance the mass numbers and the atomic numbers in each nuclear equation.

    a. $^1_0n + {}^9_4\text{Be} \rightarrow {}^{10}_4\text{Be}$ $? = {}^{10}_4\text{Be}$

    b. $^1_0n + {}^{131}_{52}\text{Te} \rightarrow {}^{132}_{53}\text{I} + {}^0_{-1}e$ $? = {}^{132}_{53}\text{I}$

    c. $^1_0n + {}^{27}_{13}\text{Al} \rightarrow {}^{24}_{11}\text{Na} + {}^4_2\text{He}$ $? = {}^{27}_{13}\text{Al}$

    d. $^4_2\text{He} + {}^{27}_{13}\text{Al} \rightarrow {}^{30}_{15}\text{P} + {}^1_0n$ $? = {}^{30}_{15}\text{P}$

9.23   **a.**   When radiation enters the Geiger counter, it ionizes a gas in the detection tube, which produces a burst of electrical current that is detected by the instrument.

   b.   The becquerel (Bq) is the SI unit for activity. The curie (Ci) is the original unit for activity of radioactive samples.

   c.   The SI unit for absorbed dose is the gray (Gy). The rad (radiation absorbed dose) is a unit of radiation absorbed per gram of sample. It is the older unit.

   d.   A kilogray is 1000 Gy, which is equivalent to 100 000 rad.

9.25 $70.0 \text{ kg body mass} \times \dfrac{4.20 \ \mu Ci}{1 \text{ kg body mass}} = 294 \ \mu Ci \ (3 \text{ SFs})$

9.27   A half-life is the time it takes for one-half of a radioactive sample to decay.

9.29   **a.**   After one half-life, one-half of the sample would be radioactive:
   $80.0 \text{ mg} \times \frac{1}{2} = 40.0 \text{ mg} \ (3 \text{ SFs})$

   b.   After two half-lives, one-fourth of the sample would still be radioactive:
   $80.0 \text{ mg} \times \frac{1}{2} \times \frac{1}{2} = 80.0 \text{ mg} \times \frac{1}{4} = 20.0 \text{ mg} \ (3 \text{ SFs})$

   c. $18 \text{ h} \times \dfrac{1 \text{ half-life}}{6.0 \text{ h}} = 3.0 \text{ half-lives}$
   $80.0 \text{ mg} \times \frac{1}{2} \times \frac{1}{2} \times \frac{1}{2} = 80.0 \text{ mg} \times \frac{1}{8} = 10.0 \text{ mg} \ (3 \text{ SFs})$

   d. $24 \text{ h} \times \dfrac{1 \text{ half-life}}{6.0 \text{ h}} = 4.0 \text{ half-lives}$
   $80.0 \text{ mg} \times \frac{1}{2} \times \frac{1}{2} \times \frac{1}{2} \times \frac{1}{2} = 80.0 \text{ mg} \times \frac{1}{16} = 5.00 \text{ mg} \ (3 \text{ SFs})$

9.31   The radiation level in a radioactive sample is cut in half with each half-life; the half-life of Sr-85 is 65 days.

   For the radiation level to drop to one-fourth of its original level, $\frac{1}{4} = \frac{1}{2} \times \frac{1}{2}$ or 2 half-lives

   $2 \text{ half-lives} \times \dfrac{65 \text{ days}}{1 \text{ half-life}} = 130 \text{ days} \ (2 \text{ SFs})$

   For the radiation level to drop to one-eighth of its original level, $\frac{1}{8} = \frac{1}{2} \times \frac{1}{2} \times \frac{1}{2}$ or 3 half-lives

   $3 \text{ half-lives} \times \dfrac{65 \text{ days}}{1 \text{ half-life}} = 195 \text{ days} \ (2 \text{ SFs})$

9.33   **a.**   Since the elements calcium and phosphorus are part of the bone, any calcium or phosphorus atom, regardless of isotope, will be carried to and become part of the bony structures of the body. Once there, the radiation emitted by any radioactive isotope can be used for diagnosis or treatment of bone diseases.

   b. $^{89}_{38}\text{Sr} \rightarrow \ ^{89}_{39}\text{Y} + \ ^{0}_{-1}e$

   Strontium (Sr) acts much like calcium (Ca) because both are Group 2A (2) elements. The body will accumulate radioactive strontium in bones in the same way that it incorporates calcium. Once the strontium isotope is absorbed by the bone, the beta radiation will destroy cancer cells.

9.35 $4.0 \text{ mL solution} \times \dfrac{45 \ \mu Ci}{1 \text{ mL solution}} = 180 \ \mu Ci \text{ of selenium-75} \ (2 \text{ SFs})$

9.37   Nuclear fission is the splitting of a large atom into smaller fragments with a simultaneous release of large amounts of energy.

9.39 $^1_0n + ^{235}_{92}U \rightarrow ^{131}_{50}Sn + ^{103}_{42}Mo + 2^1_0n + \text{energy}$ $? = ^{103}_{42}Mo$

9.41 **a.** Neutrons bombard a nucleus in the fission process.
　　　b. The nuclear process that occurs in the Sun is fusion.
　　　c. Fission is the process where a large nucleus splits into smaller nuclei.
　　　d. Fusion is the process where small nuclei combine to form larger nuclei.

9.43 **a.** $^{11}_6C$
　　　b.

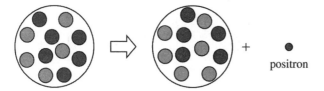

9.45

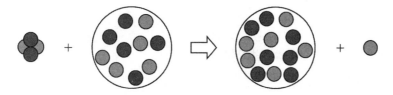

9.47 Half of a radioactive sample decays with each half-life:

$$6.4 \ \mu Ci \text{ of } ^{14}_6C \xrightarrow{1 \text{ half-life}} 3.2 \ \mu Ci \text{ of } ^{14}_6C \xrightarrow{1 \text{ half-life}} 1.6 \ \mu Ci \text{ of } ^{14}_6C \xrightarrow{1 \text{ half-life}} 0.80 \ \mu Ci \text{ of } ^{14}_6C$$

$\therefore$ the activity of carbon-14 drops to 0.80 μCi in 3 half-lives or 3×5730 years, which makes the age of the painting 17 200 years.

9.49 **a.** $^{25}_{11}Na$ has 11 protons and $25 - 11 = 14$ neutrons.
　　　b. $^{61}_{28}Ni$ has 28 protons and $61 - 28 = 33$ neutrons.
　　　c. $^{84}_{37}Rb$ has 37 protons and $84 - 37 = 47$ neutrons.
　　　d. $^{110}_{47}Ag$ has 47 protons and $110 - 47 = 63$ neutrons.

9.51 **a.** In alpha decay, a helium nucleus is emitted from the nucleus of a radioisotope. In beta decay, a neutron in an unstable nucleus decays to form a proton and an electron, which is emitted as a beta particle. In gamma emission, high-energy radiation is emitted from the nucleus of a radio-isotope.
　　　b. Alpha radiation is symbolized by α or ^4_2He. Beta radiation is symbolized by β or $^0_{-1}e$. Gamma radiation is symbolized by γ or $^0_0\gamma$.

9.53 **a.** gamma emission
　　　b. positron emission
　　　c. alpha decay

9.55 **a.** $^{225}_{90}\text{Th} \rightarrow ^{221}_{88}\text{Ra} + ^{4}_{2}\text{He}$

   b. $^{210}_{83}\text{Bi} \rightarrow ^{206}_{81}\text{Tl} + ^{4}_{2}\text{He}$

   c. $^{137}_{55}\text{Cs} \rightarrow ^{137}_{56}\text{Ba} + ^{0}_{-1}e$

   d. $^{126}_{50}\text{Sn} \rightarrow ^{126}_{51}\text{Sb} + ^{0}_{-1}e$

   e. $^{18}_{9}\text{F} \rightarrow ^{18}_{8}\text{O} + ^{0}_{+1}e$

9.57 **a.** $^{4}_{2}\text{He} + ^{14}_{7}\text{N} \rightarrow ^{17}_{8}\text{O} + ^{1}_{1}\text{H}$ $? = ^{17}_{8}\text{O}$

   b. $^{4}_{2}\text{He} + ^{27}_{13}\text{Al} \rightarrow ^{30}_{14}\text{Si} + ^{1}_{1}\text{H}$ $? = ^{1}_{1}\text{H}$

   c. $^{1}_{0}n + ^{235}_{92}\text{U} \rightarrow ^{90}_{38}\text{Sr} + 3^{1}_{0}n + ^{143}_{54}\text{Xe}$ $? = ^{143}_{54}\text{Xe}$

9.59 **a.** $^{16}_{8}\text{O} + ^{16}_{8}\text{O} \rightarrow ^{28}_{14}\text{Si} + ^{4}_{2}\text{He}$

   b. $^{18}_{8}\text{O} + ^{249}_{98}\text{Cf} \rightarrow ^{263}_{106}\text{Sg} + 4^{1}_{0}n$

   c. $^{222}_{86}\text{Rn} \rightarrow ^{218}_{84}\text{Po} + ^{4}_{2}\text{He}$

9.61 Half of a radioactive sample decays with each half-life:

   $1.2 \text{ mg of } ^{32}_{15}\text{P} \xrightarrow{\text{1 half-life}} 0.60 \text{ mg of } ^{32}_{15}\text{P} \xrightarrow{\text{1 half-life}} 0.30 \text{ mg of } ^{32}_{15}\text{P}$

   $\therefore$ 2 half-lives must have elapsed during this time (28 days), yielding the half-life for phosphorus-32:

   $$\frac{28 \text{ days}}{2 \text{ half-lives}} = 14 \text{ days/half-life (2 SFs)}$$

9.63 **a.** $^{47}_{20}\text{Ca} \rightarrow ^{47}_{21}\text{Sc} + ^{0}_{-1}e$

   b. First, calculate the number of half-lives that have passed:

   $$18 \text{ days} \times \frac{1 \text{ half-life}}{4.5 \text{ days}} = 4.0 \text{ half-lives}$$

   Now we can calculate the number of milligrams of calcium-47 that remain:

   $$16 \text{ mg} \times \tfrac{1}{2} \times \tfrac{1}{2} \times \tfrac{1}{2} \times \tfrac{1}{2} = 16 \text{ mg} \times \tfrac{1}{16} = 1.0 \text{ mg of calcium-47 (2 SFs)}$$

   c. Half of a radioactive sample decays with each half-life:

   $4.8 \text{ mg of } ^{47}_{20}\text{Ca} \xrightarrow{\text{1 half-life}} 2.4 \text{ mg of } ^{47}_{20}\text{Ca} \xrightarrow{\text{1 half-life}} 1.2 \text{ mg of } ^{47}_{20}\text{Ca}$

   $\therefore$ 2 half-lives have passed.

   $$2 \text{ half-lives} \times \frac{4.5 \text{ d}}{1 \text{ half-life}} = 9.0 \text{ days (2 SFs)}$$

9.65 First, calculate the number of half-lives that have passed since the technician was exposed:

$$36 \cancel{h} \times \frac{1 \text{ half-life}}{12 \cancel{h}} = 3.0 \text{ half-lives}$$

Because the activity of a radioactive sample is cut in half with each half-life, the activity must have been double its present value before each half-life. For 3.0 half-lives, we need to double the value 3 times:

$$2.0 \ \mu\text{Ci of } {}^{42}_{19}\text{K} \xleftarrow{\text{1 half-life}} 4.0 \ \mu\text{Ci of } {}^{42}_{19}\text{K} \xleftarrow{\text{1 half-life}} 8.0 \ \mu\text{Ci of } {}^{42}_{19}\text{K} \xleftarrow{\text{1 half-life}} 16 \ \mu\text{Ci of } {}^{42}_{19}\text{K}$$

$$2.0 \ \mu\text{Ci} \times (2 \times 2 \times 2) = 16 \ \mu\text{Ci (2 SFs)}$$

9.67 First, calculate the number of half-lives that have passed:

$$24 \cancel{h} \times \frac{1 \text{ half-life}}{6.0 \cancel{h}} = 4.0 \text{ half-lives}$$

Now we can calculate the number of milligrams of technicium-99m that remain:

$$120 \text{ mg} \times \tfrac{1}{2} \times \tfrac{1}{2} \times \tfrac{1}{2} \times \tfrac{1}{2} = 120 \text{ mg} \times \tfrac{1}{16} = 7.5 \text{ mg (2 SFs)}$$

9.69 The irradiation of meats, fruits, and vegetables kills bacteria such as *E. coli* that can cause food-borne illnesses. In addition, food spoilage is deterred and shelf life is extended.

9.71 In the fission process, an atom splits into smaller nuclei with a simultaneous release of large amounts of energy. In fusion, two (or more) small nuclei combine (fuse) to form a larger nucleus, with a simultaneous release of large amounts of energy.

9.73 Fusion occurs naturally in the Sun and other stars.

9.75 **a.** $\quad {}^{238}_{92}\text{U} \rightarrow {}^{234}_{90}\text{Th} + {}^{4}_{2}\text{He} \qquad\qquad ? = {}^{4}_{2}\text{He}$

       b. $\quad {}^{234}_{90}\text{Th} \rightarrow {}^{234}_{91}\text{Pa} + {}^{0}_{-1}e \qquad\qquad ? = {}^{234}_{91}\text{Pa}$

       c. $\quad {}^{226}_{88}\text{Ra} \rightarrow {}^{222}_{86}\text{Rn} + {}^{4}_{2}\text{He} \qquad\qquad ? = {}^{226}_{88}\text{Ra}$

9.77 **a.** $\quad {}^{23m}_{12}\text{Mg} \rightarrow {}^{23}_{12}\text{Mg} + {}^{0}_{0}\gamma \qquad\qquad ? = {}^{23}_{12}\text{Mg}$

       b. $\quad {}^{61}_{30}\text{Zn} \rightarrow {}^{61}_{29}\text{Cu} + {}^{0}_{+1}e \qquad\qquad ? = {}^{0}_{+1}e$

       c. $\quad {}^{12}_{6}\text{C} + {}^{249}_{98}\text{Cf} \rightarrow {}^{257}_{104}\text{Rf} + 4{}^{1}_{0}n \qquad\qquad ? = {}^{257}_{104}\text{Rf}$

9.79 First, calculate the number of half-lives that have elapsed:

$$18 \cancel{\text{days}} \times \frac{1 \text{ half-life}}{4.5 \cancel{\text{days}}} = 4.0 \text{ half-lives}$$

Because the activity of a radioactive sample is cut in half with each half-life, the activity must have been double its present value before each half-life. For 4.0 half-lives, we need to double the value 4 times:

$$4.0 \ \mu\text{Ci of } {}^{47}_{20}\text{Ca} \xleftarrow{\text{1 half-life}} 8.0 \ \mu\text{Ci of } {}^{47}_{20}\text{Ca} \xleftarrow{\text{1 half-life}} 16 \ \mu\text{Ci of } {}^{47}_{20}\text{Ca} \xleftarrow{\text{1 half-life}} 32 \mu\text{Ci of } {}^{47}_{20}\text{Ca}$$

$$\xleftarrow{\text{1 half-life}} 64 \ \mu\text{Ci of } {}^{47}_{20}\text{Ca}$$

$$4.0 \ \mu\text{Ci} \times (2 \times 2 \times 2 \times 2) = 64 \ \mu\text{Ci (2 SFs)}$$

Introduction to Organic Chemistry: Alkanes

10.1 Organic compounds contain C and H and sometimes O, N, or a halogen atom. Inorganic compounds usually contain elements other than C and H.
  a. KCl is inorganic.
  b. C_4H_{10} is organic.
  c. C_2H_6O is organic.
  d. H_2SO_4 is inorganic.
  e. $CaCl_2$ is inorganic.
  f. C_3H_7Cl is organic.

10.3 **a.** Inorganic compounds are usually soluble in water.
  b. Organic compounds have lower boiling points than most inorganic compounds.
  c. Organic compounds contain carbon and hydrogen.
  d. Inorganic compounds contain ionic bonds.

10.5 **a.** Ethane boils at −89 °C.
  b. Ethane burns vigorously in air.
  c. NaBr is a solid at 250 °C.
  d. NaBr dissolves in water.

10.7 VSEPR theory predicts that the four bonds between carbon and hydrogen in CH_4 will be as far apart as possible, which means that the hydrogen atoms are at the corners of a tetrahedron.

10.9 **a.** Pentane has a carbon chain of five (5) carbon atoms.
  b. Ethane has a carbon chain of two (2) carbon atoms.
  c. Hexane has a carbon chain of six (6) carbon atoms.

10.11 **a.** CH_4
  b. $CH_3—CH_3$
  c. $CH_3—CH_2—CH_2—CH_2—CH_3$
  d. △

10.13 Two structures are isomers if they have the same molecular formula, but different arrangements of atoms.
  a. These condensed structural formulas represent the same molecule; the only difference is due to rotation of the structure. Each has a $CH_3—$ group attached to the middle carbon in a three-carbon chain.
  b. The molecular formula of both these condensed structural formulas is C_5H_{12}. However, they represent isomers because the C atoms are bonded in a different order; they have different arrangements. In the first, there is a $CH_3—$ group attached to carbon 2 of a four-carbon chain, and in the other, there is a five-carbon chain.
  c. The molecular formula of both these condensed structural formulas is C_6H_{14}. However, they represent isomers because the C atoms are bonded in a different order; they have different arrangements. In the first, there is a $CH_3—$ group attached to carbon 3 of a five-carbon chain, and in the other, there is a $CH_3—$ group on carbon 2 and carbon 3 of a four-carbon chain.

10.15 **a.** 2,2-dimethylpropane
       b. 2,3-dimethylpentane
       c. 4-ethyl-2,2-dimethylhexane
       d. methylcyclopentane
       e. 1-fluoropropane
       f. 2-chloropropane

10.17 Draw the main chain with the number of carbon atoms in the ending of the name. For example, butane has a main chain of four carbon atoms, and hexane has a main chain of six carbon atoms. Attach substituents on the carbon atoms indicated. For example, in 3-methylpentane, a CH_3- group is bonded to carbon 3 of a five-carbon chain.

                      CH_3
                      |
       a. 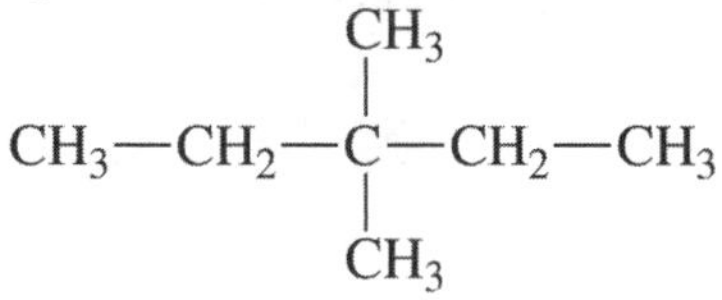 $CH_3-CH_2-C-CH_2-CH_3$
                      |
                      CH_3

            CH_3 CH_3 CH_3
            | | |
       b. $CH_3-CH-CH-CH_2-CH-CH_3$

            CH_3 CH_2-CH_3 CH_3
            | | |
       c. $CH_3-CH-CH-CH_2-CH-CH_2-CH_2-CH_3$
       d. $Br-CH_2-CH_2-Cl$

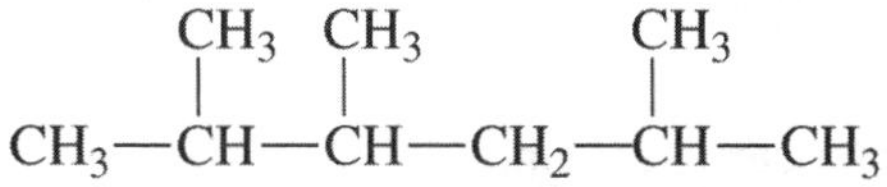

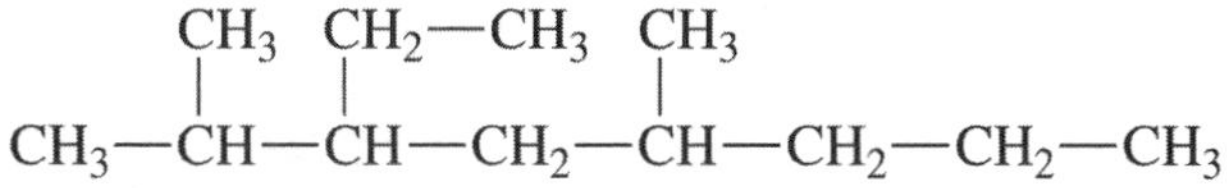

              Br   Cl
              | |
       e. 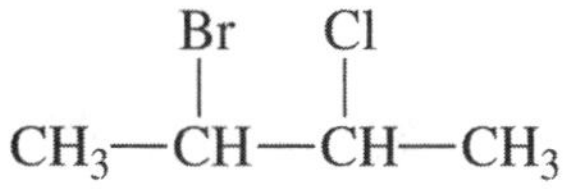 $CH_3-CH-CH-CH_3$

10.19 **a.**

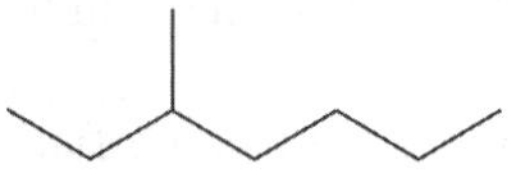

       b.

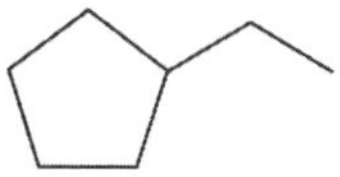

       c.

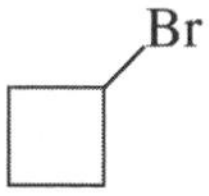

       d.

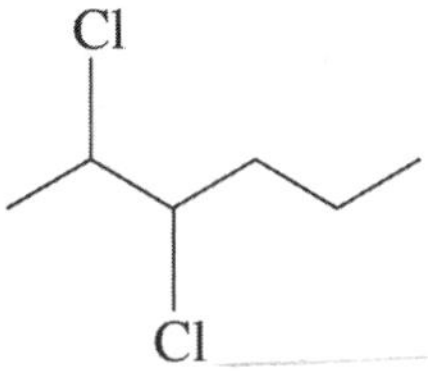

10.21 **a.** $CH_3-CH_2-CH_2-CH_2-CH_2-CH_2-CH_3$
       b. Heptane is a liquid at room temperature since it has a boiling point of 98 °C.
       c. Heptane is nonpolar, which makes it insoluble in water.
       d. Since the density of heptane (0.68 g/mL) is less than that of water, heptane will float on water.
       e. $C_7H_{16}(g) + 11O_2(g) \xrightarrow{\Delta} 7CO_2(g) + 8H_2O(g) + energy$

10.23 In combustion, a hydrocarbon reacts with oxygen to yield CO_2 and H_2O.

    a. $2C_2H_6(g) + 7O_2(g) \xrightarrow{\Delta} 4CO_2(g) + 6H_2O(g) + energy$

    b. $2C_3H_6(g) + 9O_2(g) \xrightarrow{\Delta} 6CO_2(g) + 6H_2O(g) + energy$

    c. $2C_8H_{18}(g) + 25O_2(g) \xrightarrow{\Delta} 16CO_2(g) + 18H_2O(g) + energy$

10.25   **a.** Alcohols contain a hydroxyl group (—OH) attached to a carbon chain.
    b. Alkenes have carbon–carbon double bonds.
    c. Aldehydes contain a carbonyl group (C=O) attached to at least one hydrogen atom.
    d. Esters contain a carboxyl group (—COO—) attached to two carbon atoms.

10.27   **a.** Ethers have an —O— group.
    b. Alcohols have an —OH group.
    c. Ketones have a C=O group between alkyl groups.
    d. Carboxylic acids have a —COOH group.
    e. Amines contain an N atom bonded to at least one carbon atom.

10.29   **a.** ether, aromatic, alcohol, ketone
    b. ether, aromatic, alkene, ester

10.31   ether, aromatic, amide

10.33   **a.** Organic compounds have mostly covalent bonds; inorganic compounds have ionic as well as polar covalent bonds, and a few have nonpolar covalent bonds.
    b. Most organic compounds are insoluble in water; many inorganic compounds are soluble in water.
    c. Most organic compounds have low melting points; inorganic compounds have high melting points.
    d. Most organic compounds are flammable; inorganic compounds are not usually flammable.

10.35   **a.** Butane melts at $-138\ °C$.
    b. Butane burns vigorously in air.
    c. KCl melts at $770\ °C$.
    d. KCl contains ionic bonds.
    e. Butane is a gas at room temperature.

10.37   **a.** aldehyde, aromatic
    b. alkene, aldehyde, aromatic
    c. ketone

10.39   amine, carboxylic acid, amide, aromatic, ester

10.41   **a.** 2,2-dimethylbutane
    b. chloroethane
    c. 2-bromo-4-ethylhexane
    d. cyclohexane

10.43 **a.**

$$CH_3-CH_2-\underset{\underset{\displaystyle CH_2-CH_3}{\displaystyle |}}{CH}-CH_2-CH_2-CH_3$$

b.

$$CH_3-\underset{\underset{\displaystyle CH_3}{\displaystyle |}}{CH}-\underset{\underset{\displaystyle CH_3}{\displaystyle |}}{CH}-CH_2-CH_3$$

c.

$$Cl-CH_2-CH_2-\underset{\underset{\displaystyle Cl}{\displaystyle |}}{\overset{\overset{\displaystyle CH_3}{\displaystyle |}}{C}}-CH_2-CH_2-CH_2-CH_3$$

10.45 **a.** $C_5H_{12}(g) + 8O_2(g) \xrightarrow{\Delta} 5CO_2(g) + 6H_2O(g) + energy$

b. $C_4H_8(g) + 6O_2(g) \xrightarrow{\Delta} 4CO_2(g) + 4H_2O(g) + energy$

c. $2C_6H_{14}(g) + 19O_2(g) \xrightarrow{\Delta} 12CO_2(g) + 14H_2O(g) + energy$

d. $2C_3H_6(g) + 9O_2(g) \xrightarrow{\Delta} 6CO_2(g) + 6H_2O(g) + energy$

10.47 **a.** amine
b. alcohol
c. ester
d. alkene

10.49 **a.** alcohol
b. alkene
c. aldehyde
d. alkane
e. carboxylic acid
f. amine

10.51 **a.** $C_3H_8(g) + 5O_2(g) \xrightarrow{\Delta} 3CO_2(g) + 4H_2O(g) + energy$

b. $5.0 \text{ lb } C_3H_8 \times \dfrac{454 \text{ g } C_3H_8}{1 \text{ lb } C_3H_8} \times \dfrac{1 \text{ mole } C_3H_8}{44.1 \text{ g } C_3H_8} \times \dfrac{3 \text{ moles } CO_2}{1 \text{ mole } C_3H_8} \times \dfrac{44.0 \text{ g } CO_2}{1 \text{ mole } CO_2} \times \dfrac{1 \text{ kg } CO_2}{1000 \text{ g } CO_2}$

$= 6.8 \text{ kg of } CO_2 \ (2 \text{ SFs})$

10.53 **a.** $C_5H_{12}(g) + 8O_2(g) \xrightarrow{\Delta} 5CO_2(g) + 6H_2O(g) + energy$

b. Molar mass of pentane $(C_5H_{12}) = 5(12.0 \text{ g}) + 12(1.01 \text{ g}) = 72.1 \text{ g/mole}$

c. $1 \text{ gal} \times \dfrac{4 \text{ qt}}{1 \text{ gal}} \times \dfrac{946 \text{ mL}}{1 \text{ qt}} \times \dfrac{0.63 \text{ g}}{1 \text{ mL}} \times \dfrac{1 \text{ mole } C_5H_{12}}{72.1 \text{ g } C_5H_{12}} \times \dfrac{845 \text{ kcal}}{1 \text{ mole } C_5H_{12}} = 2.8 \times 10^4 \text{ kcal} \ (2 \text{ SFs})$

d. $1 \text{ gal} \times \dfrac{4 \text{ qt}}{1 \text{ gal}} \times \dfrac{946 \text{ mL}}{1 \text{ qt}} \times \dfrac{0.63 \text{ g}}{1 \text{ mL}} \times \dfrac{1 \text{ mole } C_5H_{12}}{72.1 \text{ g } C_5H_{12}} \times \dfrac{5 \text{ moles } CO_2}{1 \text{ mole } C_5H_{12}} \times \dfrac{22.4 \text{ L } CO_2 (STP)}{1 \text{ mole } CO_2}$

$= 3.7 \times 10^3 \text{ L of } CO_2 \text{ at STP} \ (2 \text{ SFs})$

10.55

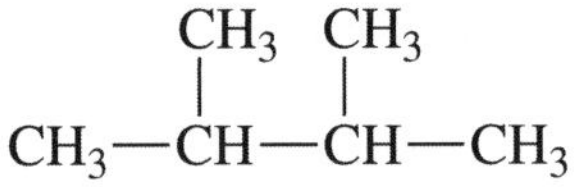

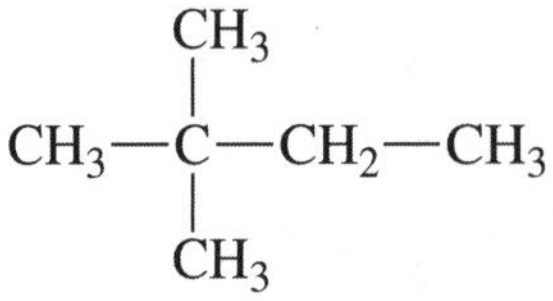

$$CH_3-CH-CH-CH_3$$

with CH_3 groups on the two middle carbons.

$$CH_3-C-CH_2-CH_3$$

with two CH_3 groups above and one below the second carbon.

10.57 **a.** $CH_3-CH_2-CH_3$

b. $C_3H_8(g) + 5O_2(g) \xrightarrow{\Delta} 3CO_2(g) + 4H_2O(g) + \text{energy}$

c. $12.0 \text{ L } C_3H_8 \times \dfrac{1 \text{ mole } C_3H_8}{22.4 \text{ L } C_3H_8 \text{ (STP)}} \times \dfrac{5 \text{ moles } O_2}{1 \text{ mole } C_3H_8} \times \dfrac{32.0 \text{ g } O_2}{1 \text{ mole } O_2} = 85.7 \text{ g of } O_2 \ (3 \text{ SFs})$

d. $12.0 \text{ L } C_3H_8 \times \dfrac{1 \text{ mole } C_3H_8}{22.4 \text{ L } C_3H_8 \text{ (STP)}} \times \dfrac{3 \text{ moles } CO_2}{1 \text{ mole } C_3H_8} \times \dfrac{44.0 \text{ g } CO_2}{1 \text{ mole } CO_2} = 70.7 \text{ g of } CO_2 \ (3 \text{ SFs})$

10.59 **a.** 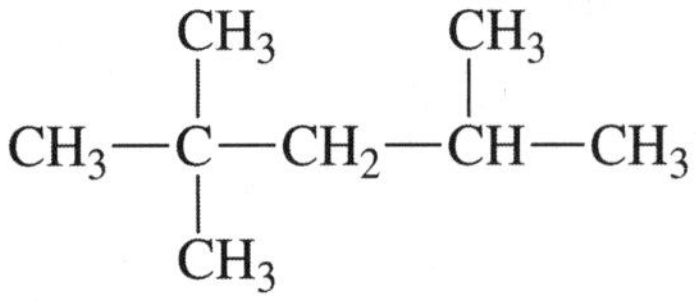 $CH_3-C-CH_2-CH-CH_3$ with two CH_3 groups (above and below the second carbon) and one CH_3 above the fourth carbon.   2,2,4-Trimethylpentane

b. The molecular formula is C_8H_{18}.

c. $2C_8H_{18}(g) + 25O_2(g) \xrightarrow{\Delta} 16CO_2(g) + 18H_2O(g) + \text{energy}$

11

Unsaturated Hydrocarbons

11.1 **a.** A condensed structural formula with a carbon–carbon double bond is an alkene.
b. A condensed structural formula with a carbon–carbon triple bond is an alkyne.
c. A condensed structural formula with a carbon–carbon double bond in a ring is a cycloalkene.
d. A skeletal formula with a carbon–carbon double bond is an alkene.

11.3 Propene contains a double bond, and propyne has a triple bond. Propene has six hydrogen atoms, and propyne has only four hydrogen atoms.

11.5 **a.** The two-carbon compound with a double bond is named ethene.
b. This alkene has a three-carbon chain with a methyl substituent. The name is methylpropene.
c. This alkyne has a five-carbon chain with the triple bond between carbon 2 and carbon 3. The name is 2-pentyne.
d. This is a four-carbon cyclic structure with a double bond. The name is cyclobutene.

11.7 **a.** Propene is the three-carbon alkene.

$H_2C\!=\!CH\!-\!CH_3$

b. 1-Pentene is the five-carbon compound with a double bond between carbon 1 and carbon 2.

$H_2C\!=\!CH\!-\!CH_2\!-\!CH_2\!-\!CH_3$

c. 2-Methyl-1-butene has a four-carbon chain with a double bond between carbon 1 and carbon 2 and a methyl group attached to carbon 2.

$$\begin{array}{c} \quad\;\; CH_3 \\ \quad\;\; | \\ H_2C\!=\!C\!-\!CH_2\!-\!CH_3 \end{array}$$

d. Cyclohexene is a six-carbon cyclic compound with a double bond.

e. 1-Butyne is a four-carbon compound with a triple bond between carbon 1 and carbon 2.

$HC\!\equiv\!C\!-\!CH_2\!-\!CH_3$

f. 1-Bromo-3-hexyne is a six-carbon compound with a triple bond between carbon 3 and carbon 4 and a bromine atom bonded to carbon 1.

$Br\!-\!CH_2\!-\!CH_2\!-\!C\!\equiv\!C\!-\!CH_2\!-\!CH_3$

11.9 **a.** *cis*-2-Butene; this is a four-carbon compound with a double bond between carbon 2 and carbon 3. Both methyl groups are on the same side of the double bond; it is the cis isomer.
b. *trans*-3-Octene; this compound has eight carbons with a double bond between carbon 3 and carbon 4. The alkyl groups are on the opposite sides of the double bond; it is the trans isomer.
c. *cis*-3-Heptene; this is a seven-carbon compound with a double bond between carbon 3 and carbon 4. Both alkyl groups are on the same side of the double bond; it is the cis isomer.

101

11.11 **a.** *trans*-1-Bromo-2-chloroethene has a two-carbon chain with a double bond. The trans isomer has the attached groups on the opposite sides of the double bond.

$$\underset{H}{\overset{Br}{\diagdown}}C=C\underset{Cl}{\overset{H}{\diagup}}$$

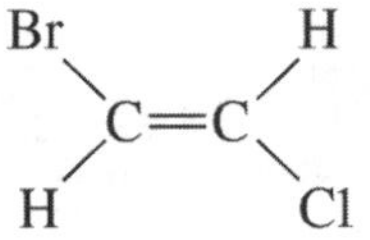

b. *cis*-2-Pentene has a five-carbon chain with a double bond between carbon 2 and carbon 3. The cis isomer has the alkyl groups on the same side of the double bond.

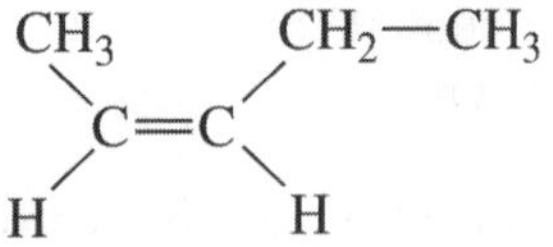

$$\underset{H}{\overset{CH_3}{\diagdown}}C=C\underset{H}{\overset{CH_2-CH_3}{\diagup}}$$

c. *trans*-3-Heptene has a seven-carbon chain with a double bond between carbon 3 and carbon 4. The trans isomer has the alkyl groups on the opposite sides of the double bond.

$$\underset{H}{\overset{CH_3-CH_2}{\diagdown}}C=C\underset{CH_2-CH_2-CH_3}{\overset{H}{\diagup}}$$

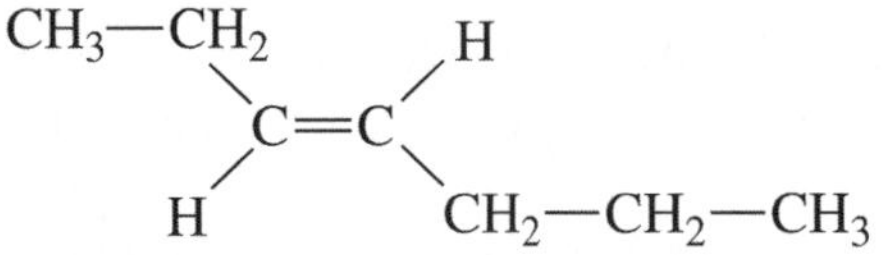

11.13 **a.** Hydrogenation of an alkene gives the saturated compound, the alkane.

$$CH_3-CH_2-CH_2-CH_2-CH_3$$

b. In a hydration reaction, the H— and —OH from water (HOH) add to the carbon atoms in the double bond to form an alcohol.

$$\overset{\displaystyle OH}{\underset{\displaystyle |}{CH_3-CH-CH_2-CH_3}}$$

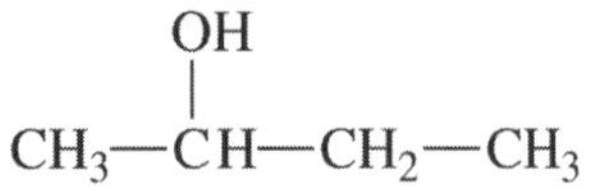

c. The H— and —OH from water (HOH) add to the carbon atoms in the double bond to form an alcohol. Hydration of a cycloalkene forms a cycloalkanol.

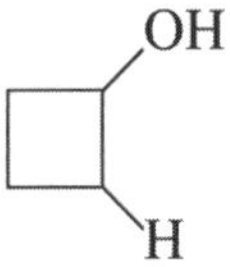

d. When Cl_2 is added to a cycloalkene, the chlorine atoms add to the carbon atoms in the double bond, producing a dichlorocycloalkane.

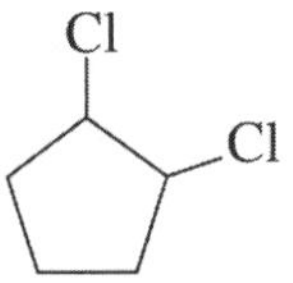

e. Complete hydrogenation of an alkyne gives the saturated compound, the alkane.

$$CH_3-CH_2-CH_2-CH_2-CH_3$$

11.15 A polymer is a very large molecule composed of small units that are repeated many times.

11.17 $3\ H_2C=\overset{\displaystyle CH_3}{\underset{\displaystyle |}{CH}} \longrightarrow$

$$-\overset{H}{\underset{H}{C}}-\overset{CH_3}{\underset{H}{C}}-\overset{H}{\underset{H}{C}}-\overset{CH_3}{\underset{H}{C}}-\overset{H}{\underset{H}{C}}-\overset{CH_3}{\underset{H}{C}}-$$

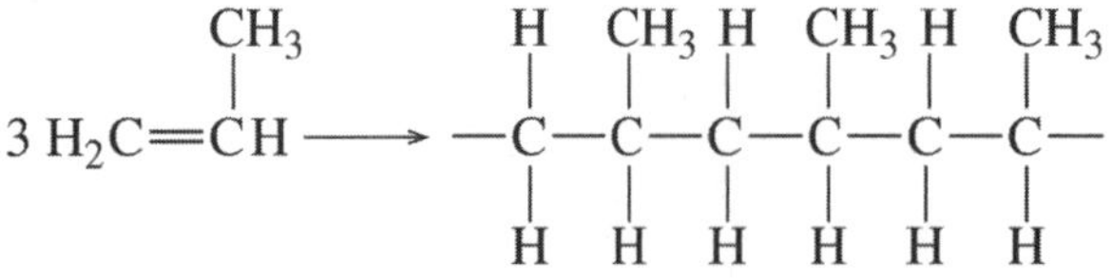

11.19 Cyclohexane, C_6H_{12}, is a cycloalkane with 6 carbon atoms and 12 hydrogen atoms. The carbon atoms are connected in a ring by single bonds. Benzene, C_6H_6, is an aromatic compound with 6 carbon atoms and 6 hydrogen atoms. The carbon atoms are connected in a ring where the electrons are equally shared among the six carbon atoms.

11.21 Aromatic compounds that contain a benzene ring with a single substituent are usually named as benzene derivatives. A benzene ring with a methyl substituent is named toluene. The methyl group is attached to carbon 1 and the ring is numbered to give the lower numbers to other substituents.
   a. 2-chlorotoluene
   b. ethylbenzene
   c. 1,3,5-trichlorobenzene

11.23 a.

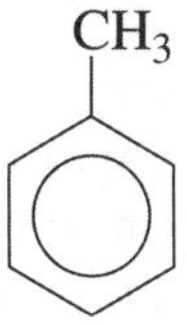

b.

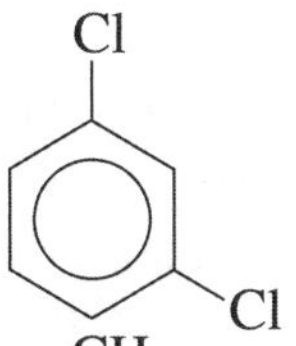

c.

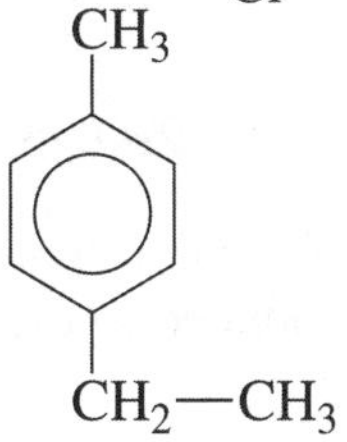

11.25

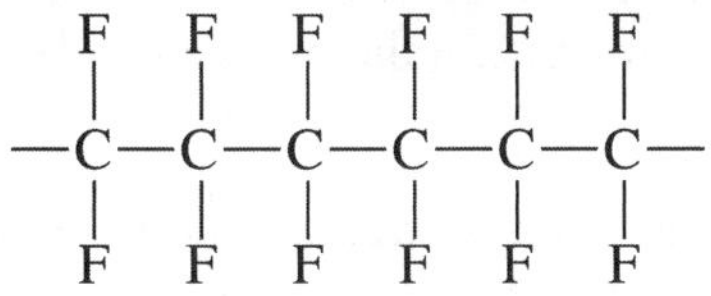

11.27 a.

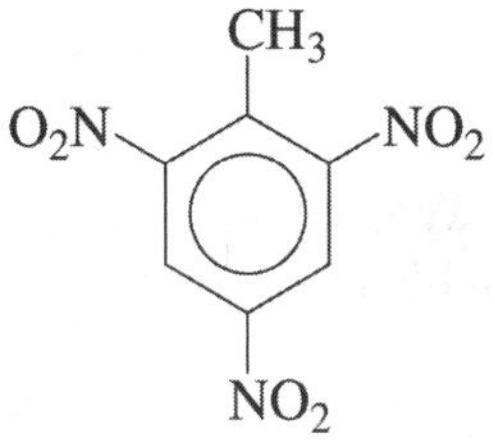

b.

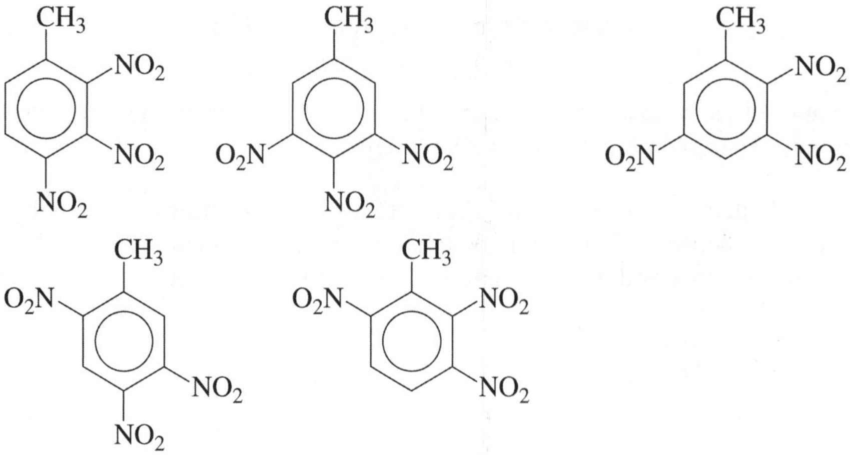

CH₃ structures with NO₂, O₂N substituents on benzene rings (structural isomers of trinitrotoluene)

11.29 All the compounds have three carbon atoms: propane has eight hydrogen atoms, cyclopropane has six hydrogen atoms, propene has six hydrogen atoms, and propyne has four hydrogen atoms. Propane is a saturated alkane, and cyclopropane is a saturated cyclic hydrocarbon. Both propene and propyne are unsaturated hydrocarbons, but propene has a double bond and propyne has a triple bond.

11.31 **a.** This alkene has a five-carbon chain with a double bond between carbon 1 and carbon 2 and a methyl group attached to carbon 2. The IUPAC name is 2-methyl-1-pentene.
   b. This alkene has a six-carbon chain with the double bond between carbon 1 and carbon 2. The substituents are a bromine atom attached to carbon 4 and a methyl group on carbon 5. The IUPAC name is 4-bromo-5-methyl-1-hexene.
   c. This is a five-carbon cyclic structure with a double bond. The name is cyclopentene.

11.33 **a.** These structures represent a pair of structural isomers. In one isomer, the chlorine is attached to one of the carbons in the double bond; in the other isomer, the carbon bonded to the chlorine is not part of the double bond.
   b. These structures are cis–trans isomers. In the cis isomer, the two methyl groups are on the same side of the double bond. In the trans isomer, the methyl groups are on the opposite sides of the double bond.

11.35 **a.** 1,1-Dibromo-2-pentyne has a five-carbon chain with a triple bond between carbon 2 and carbon 3 and two bromine atoms attached to carbon 1.

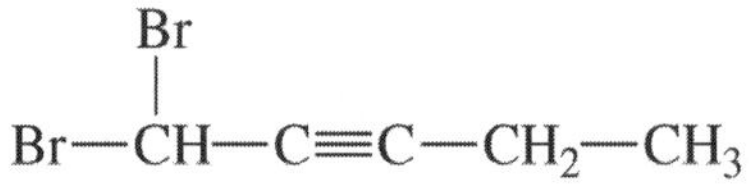

$$Br-\underset{\underset{Br}{|}}{CH}-C\equiv C-CH_2-CH_3$$

   b. *cis*-2-Heptene has a seven-carbon chain with a double bond between carbon 2 and carbon 3. In the cis isomer, both alkyl groups are on the same side of the double bond.

CH₃ and CH₂—CH₂—CH₂—CH₃ on one side; H and H below; C=C

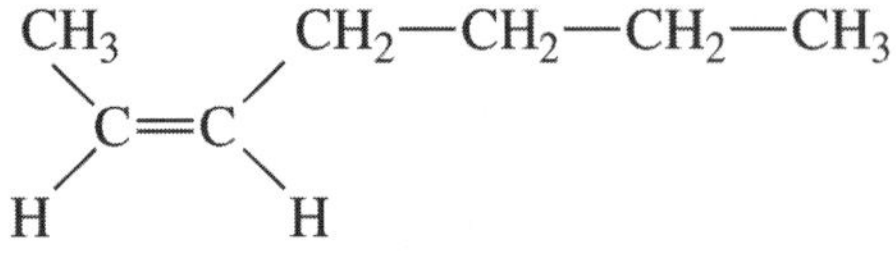

11.37 **a.**

$$CH_3 \qquad CH_2—CH_3$$
$$\diagdown \qquad \diagup$$
$$C=C$$
$$\diagup \qquad \diagdown$$
$$H \qquad \qquad H$$

cis-2-Pentene; both alkyl groups are on the same side of the double bond.

$$CH_3 \qquad \qquad H$$
$$\diagdown \qquad \diagup$$
$$C=C$$
$$\diagup \qquad \diagdown$$
$$H \qquad CH_2—CH_3$$

trans-2-Pentene; both alkyl groups are on the opposite sides of the double bond.

b.

$$CH_3—CH_2 \qquad CH_2—CH_3$$
$$\diagdown \qquad \diagup$$
$$C=C$$
$$\diagup \qquad \diagdown$$
$$H \qquad \qquad H$$

cis-3-Hexene; both alkyl groups are on the same side of the double bond.

$$CH_3—CH_2 \qquad \qquad H$$
$$\diagdown \qquad \diagup$$
$$C=C$$
$$\diagup \qquad \diagdown$$
$$H \qquad CH_2—CH_3$$

trans-3-Hexene; both alkyl groups are on the opposite sides of the double bond.

11.39 During hydrogenation, the multiple bonds are converted to single bonds and two H atoms are added for each bond converted.
   a. 3-methylpentane
   b. cyclohexane
   c. pentane

11.41 **a.** When Br_2 is added to an alkene, the bromine atoms add to the carbon atoms in the double bond, producing a dibromoalkane.

$$\qquad \quad Br \quad \; Br$$
$$\qquad \quad | \qquad |$$
$$CH_3—CH—CH—CH_3$$

   b. Hydrogenation of a cycloalkene gives the saturated compound, the cycloalkane.

   c. The H— and —OH from water (HOH) add to the carbon atoms in the double bond to form an alcohol. The H— adds to the carbon with more hydrogens (carbon 1), and the —OH bonds to the carbon with fewer hydrogen atoms (carbon 2).

$$OH$$
$$|$$

11.43

$$\quad F \quad H \quad F \quad H \quad F \quad H$$
$$\quad | \quad \; | \quad \; | \quad \; | \quad \; | \quad \; |$$
$$—C—C—C—C—C—C—$$
$$\quad | \quad \; | \quad \; | \quad \; | \quad \; | \quad \; |$$
$$\quad F \quad H \quad F \quad H \quad F \quad H$$

11.45 Aromatic compounds that contain a benzene ring with a single substituent are usually named as benzene derivatives. A benzene ring with a methyl or amino substituent is named toluene or aniline, respectively. The methyl or amino group is attached to carbon 1 and the ring is numbered to give the lower numbers to other substituents.
   a. toluene
   b. 4-bromoaniline
   c. 4-ethyltoluene

11.47   2-butene CH_3—CH=CH—CH_3

$$C_4H_8 + H_2 \rightarrow C_4H_{10}$$

Molar mass of 2-butene $(C_4H_8) = 4(12.0 \text{ g}) + 8(1.01 \text{ g}) = 56.1 \text{ g/mole}$

$$30.0 \text{ g } C_4H_8 \times \frac{1 \text{ mole } C_4H_8}{56.1 \text{ g } C_4H_8} \times \frac{1 \text{ mole } H_2}{1 \text{ mole } C_4H_8} \times \frac{2.02 \text{ g } H_2}{1 \text{ mole } H_2} = 1.08 \text{ g of } H_2 \text{ (3 SFs)}$$

11.49   Molar mass of bombykol $(C_{16}H_{30}O) = 16(12.0 \text{ g}) + 30(1.01 \text{ g}) + 1(16.00 \text{ g}) = 238.3 \text{ g/mole}$

$$50 \text{ ng} \times \frac{1 \text{ g}}{10^9 \text{ ng}} \times \frac{1 \text{ mole}}{238.3 \text{ g}} \times \frac{6.02 \times 10^{23} \text{ molecules}}{1 \text{ mole}} = 1 \times 10^{14} \text{ molecules of bombykol (1 SF)}$$

12

Organic Compounds with Oxygen and Sulfur

12.1 **a.** This compound has a two-carbon chain. The final *e* from ethane is dropped, and *ol* added to indicate an alcohol. The IUPAC name is ethanol.

  b. This compound has a four-carbon chain with a hydroxyl group attached to carbon 2. The IUPAC name is 2-butanol.

  c. This is the skeletal formula of a five-carbon chain with a hydroxyl group attached to carbon 2. The IUPAC name is 2-pentanol.

  d. This compound is a phenol because the —OH group is attached to a benzene ring. For a phenol, the carbon atom attached to the —OH is understood to be carbon 1. No number is needed to give the location of the —OH group. This compound also has a bromine atom attached to carbon 3 of the ring; the IUPAC name is 3-bromophenol.

12.3 **a.** 1-Propanol has a three-carbon chain with a hydroxyl group attached to carbon 1.

$$CH_3—CH_2—CH_2—OH$$

  b. 3-Pentanol has a five-carbon chain with a hydroxyl group attached to carbon 3.

$$\overset{\displaystyle OH}{\underset{\displaystyle |}{CH_3—CH_2—CH—CH_2—CH_3}}$$

  c. 2-Methyl-2-butanol has a four-carbon chain with a methyl group and a hydroxyl group attached to carbon 2.

$$\overset{\displaystyle OH}{\underset{\displaystyle |}{CH_3—\underset{\displaystyle |}{\underset{\displaystyle CH_3}{C}}—CH_2—CH_3}}$$

12.5 **a.** The common name of the ether with two three-carbon alkyl groups attached to an oxygen atom is dipropyl ether.

  b. The common name of the ether with a one-carbon alkyl group and a six-carbon cycloalkyl group attached to an oxygen atom is cyclohexyl methyl ether.

12.7 **a.** Ethyl methyl ether has a two-carbon group and a one-carbon group attached to oxygen by single bonds.

$$CH_3—CH_2—O—CH_3$$

  b. Cyclopropyl ethyl ether has a two-carbon group and a three-carbon cycloalkyl group attached to oxygen by single bonds.

$$CH_3—CH_2—O—\triangleleft$$

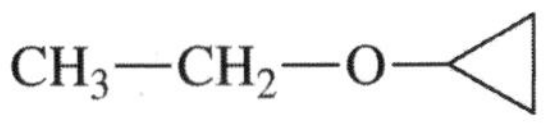

12.9 The carbon bonded to the hydroxyl group (—OH) is attached to one alkyl group in a primary (1°) alcohol, except for methanol; to two alkyl groups in a secondary alcohol (2°); and to three alkyl groups in a tertiary alcohol (3°).
 a. primary alcohol (1°)
 b. primary alcohol (1°)
 c. tertiary alcohol (3°)
 d. tertiary alcohol (3°)

12.11 a. Soluble; alcohols with one to four carbon atoms hydrogen bond with water.
 b. Slightly soluble; the water can hydrogen bond to the O in an ether.
 c. Insoluble; a carbon chain longer than four carbon atoms diminishes the effect of the —OH group.

12.13 a. Methanol has a polar —OH group that can form hydrogen bonds with water, but the alkane ethane does not.
 b. 2-Propanol is more soluble in water than 1-butanol because it has a shorter carbon chain.
 c. 1-Propanol is more soluble because it can form more hydrogen bonds with water than the ether can.

12.15 Dehydration is the removal of an H— and an —OH from adjacent carbon atoms of an alcohol to form a water molecule and the corresponding alkene.

 a. $CH_3-CH_2-CH=CH_2$

 b.

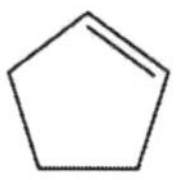

 c. $CH_3-CH_2-CH=CH-CH_3$

12.17 A primary alcohol oxidizes to an aldehyde and a secondary alcohol oxidizes to a ketone.

 a. $CH_3-CH_2-CH_2-CH_2-\overset{\overset{\displaystyle O}{\|}}{C}-H$

 b. $CH_3-\overset{\overset{\displaystyle O}{\|}}{C}-CH_2-CH_2-CH_2-CH_3$

 c.

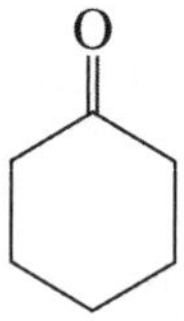

 d. $CH_3-\overset{\overset{\displaystyle O}{\|}}{C}-CH_2-\overset{\overset{\displaystyle CH_3}{|}}{C}H-CH_3$

12.19 a. acetaldehyde
 b. methyl propyl ketone
 c. formaldehyde

12.21 a. propanal
 b. 2-methyl-3-pentanone
 c. 3-methylcyclohexanone
 d. benzaldehyde

12.23 **a.** CH$_3$—C(=O)—H

b. CH$_3$—C(=O)—CH$_2$—CH$_2$—CH$_3$

c. CH$_3$—C(=O)—CH$_2$—CH$_2$—CH$_2$—CH$_3$

d. CH$_3$—CH$_2$—CH(CH$_3$)—CH$_2$—C(=O)—H

12.25 **a.** CH$_3$—C(=O)—C(=O)—CH$_2$—CH$_3$ is more soluble in water because it has two polar carbonyl groups to hydrogen bond with water.

b. Acetone is more soluble in water because it has a shorter carbon chain than 2-pentanone.

c. Propanal is more soluble in water because it has a shorter carbon chain than pentanal.

12.27 A primary alcohol will be oxidized to an aldehyde and then further oxidized to a carboxylic acid.

a. CH$_3$—CH$_2$—CH$_2$—CH$_2$—C(=O)—H ; CH$_3$—CH$_2$—CH$_2$—CH$_2$—C(=O)—OH

b. CH$_3$—CH(CH$_3$)—CH$_2$—C(=O)—H ; CH$_3$—CH(CH$_3$)—CH$_2$—C(=O)—OH

c. CH$_3$—CH$_2$—CH$_2$—C(=O)—H ; CH$_3$—CH$_2$—CH$_2$—C(=O)—OH

12.29 In reduction, an aldehyde will give a primary alcohol and a ketone will give a secondary alcohol.

a. Butyraldehyde is the four-carbon aldehyde; it will be reduced to a four-carbon primary alcohol.

CH$_3$—CH$_2$—CH$_2$—CH$_2$—OH

b. Acetone is a three-carbon ketone; it will be reduced to a three-carbon secondary alcohol.

CH$_3$—CH(OH)—CH$_3$

c. Hexanal is a six-carbon aldehyde; it reduces to a six-carbon primary alcohol.

CH$_3$—CH$_2$—CH$_2$—CH$_2$—CH$_2$—CH$_2$—OH

d. 2-Methyl-3-pentanone is a five-carbon ketone with a methyl group attached to carbon 2. It will be reduced to a five-carbon secondary alcohol with a methyl group attached to carbon 2.

CH$_3$—CH(CH$_3$)—CH(OH)—CH$_2$—CH$_3$

12.31 **a.** Achiral; there are no carbon atoms attached to four different groups.

b. Chiral; CH$_3$—CH(Br)—CH$_2$—CH$_3$ *Chiral carbon*

c. Chiral; *Chiral carbon*

$$CH_3-\overset{\overset{\displaystyle Br}{|}}{CH}-\overset{\overset{\displaystyle O}{||}}{C}-H$$

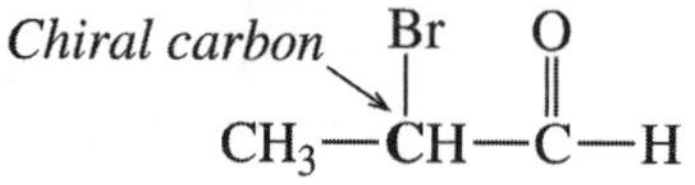

d. Achiral; there are no carbon atoms attached to four different groups.

12.33 a.

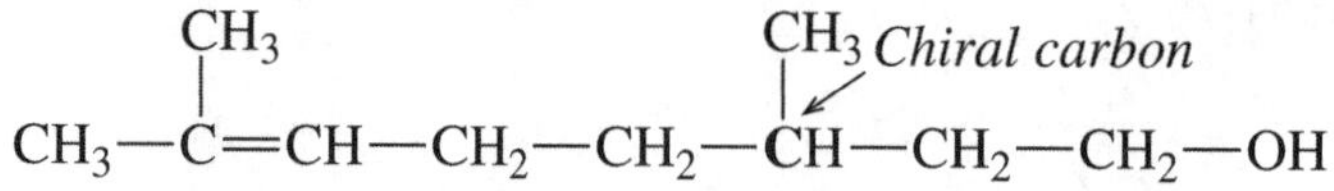

$$CH_3-\overset{\overset{\displaystyle CH_3}{|}}{C}=CH-CH_2-CH_2-\overset{\overset{\displaystyle CH_3 \,\text{\textit{Chiral carbon}}}{|}}{CH}-CH_2-CH_2-OH$$

b.

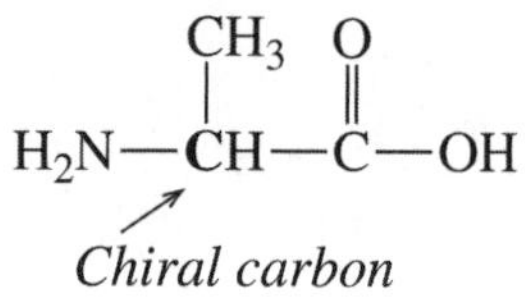

$$H_2N-\overset{\overset{\displaystyle CH_3}{|}}{CH}-\overset{\overset{\displaystyle O}{||}}{C}-OH$$

Chiral carbon

12.35 a.

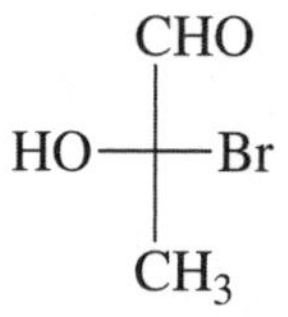

CHO
HO——Br
CH$_3$

b.

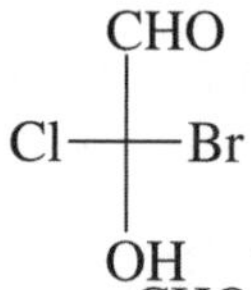

CHO
Cl——Br
OH

c.

CHO
HO——H
CH$_2$CH$_3$

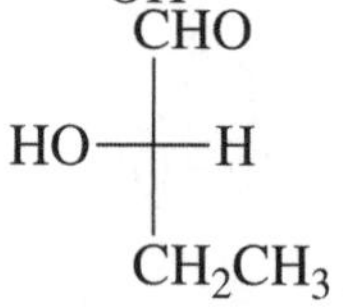

12.37 Enantiomers are nonsuperimposable mirror images.
  a. identical structures (no chiral carbons)
  b. enantiomers
  c. enantiomers
  d. enantiomers

12.39 a. phenol (aromatic + hydroxyl group)
  b. Toluene is the benzene ring attached to the methyl group.

12.41 a. phenol (aromatic + hydroxyl group)
  b. alcohol

12.43 a. 3,7-dimethyl-6-octenal
  b. The *en* in octenal signifies a double bond. The -6- indicates the double bond is between carbon 6 and carbon 7, counting from the aldehyde as carbon 1.
  c. The *al* in octenal signifies that an aldehyde group is carbon 1.
  d. $C_{10}H_{18}O(g) + 14O_2(g) \overset{\Delta}{\longrightarrow} 10CO_2(g) + 9H_2O(g) + \text{energy}$

12.45 The carbon bonded to the hydroxyl group ($-OH$) is attached to one alkyl group in a primary (1°) alcohol, except for methanol; to two alkyl groups in a secondary alcohol (2°); and to three alkyl groups in a tertiary alcohol (3°).
  a. secondary (2°) alcohol
  b. primary (1°) alcohol
  c. secondary (2°) alcohol
  d. tertiary (3°) alcohol

12.47 **a.** 2,4-dimethyl-2-pentanol
b. 4-bromo-2-pentanol
c. 3-methylphenol

12.49 **a.**

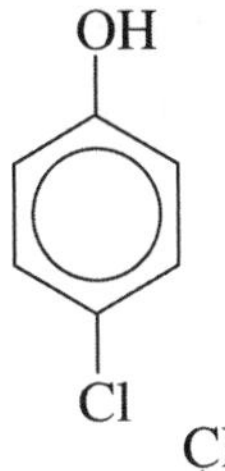

OH

Cl

b. $CH_3-\overset{\overset{\displaystyle CH_3}{|}}{CH}-\overset{\overset{\displaystyle OH}{|}}{CH}-CH_2-CH_3$

c. $CH_3-\overset{\overset{\displaystyle CH_3}{|}}{CH}-CH_2-OH$

d.

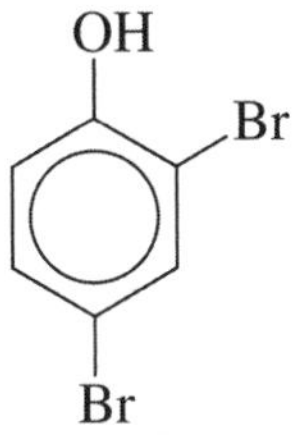

OH
Br

Br

12.51 **a.** 1-Propanol has a polar —OH group that can form hydrogen bonds with water, but the alkane butane does not.
b. 1-Propanol is more soluble because it can form more hydrogen bonds with water than the ether can.
c. Ethanol is more soluble in water than 1-hexanol because it has a shorter carbon chain.

12.53 **a.** Soluble; alcohols with one to four carbon atoms hydrogen bond with water.
b. Insoluble; a carbon chain longer than four carbon atoms diminishes the effect of the —OH group on hydrogen bonding.

12.55 **a.** Dehydration of an alcohol produces an alkene. $CH_3-CH{=}CH_2$

b. Oxidation of a primary alcohol produces an aldehyde. $CH_3-CH_2-\overset{\overset{\displaystyle O}{\|}}{C}-H$

c. Dehydration of an alcohol produces an alkene. $CH_3-CH{=}CH-CH_2-CH_3$

d. Oxidation of a secondary alcohol produces a ketone.

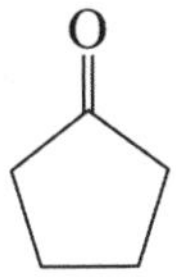

12.57 In reduction, an aldehyde will give a primary alcohol and a ketone will give a secondary alcohol.

a. $CH_3-CH_2-\overset{\overset{\displaystyle OH}{|}}{CH}-CH_3$

b.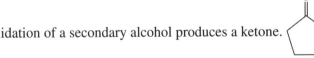

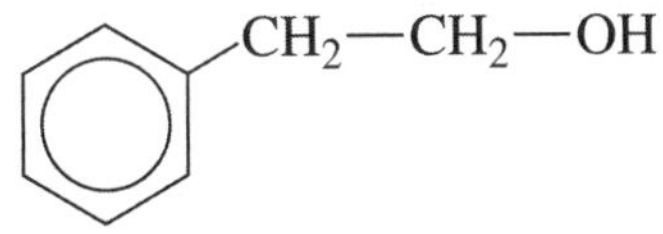

c. OH

111

12.59 **a.** 3-bromo-4-chlorobenzaldehyde
b. 3-chloropropanal
c. 2-chloro-3-pentanone

12.61 **a.** 4-Chlorobenzaldehyde is the aldehyde of benzene with a chlorine atom attached to carbon 4 of the ring.

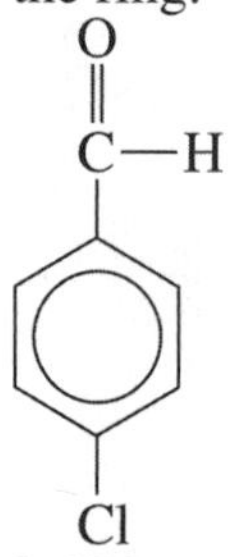

$$\underset{\displaystyle Cl}{\overset{\displaystyle \overset{O}{\overset{\|}{C}}-H}{\bigcirc}}$$

b. 3-Chloropropionaldehyde is a three-carbon aldehyde with a chlorine atom located two carbons from the carbonyl group.

$$Cl-CH_2-CH_2-\overset{\overset{O}{\|}}{C}-H$$

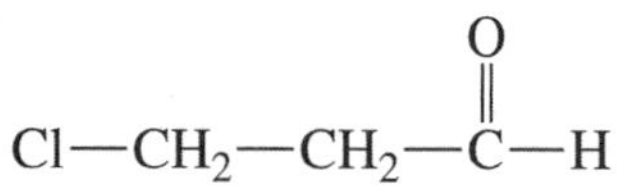

c. Ethyl methyl ketone (butanone) is a four-carbon ketone.

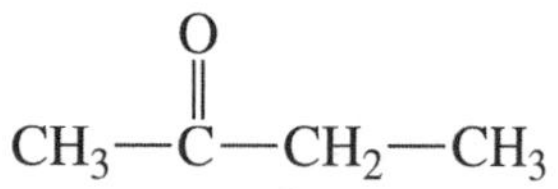

$$CH_3-\overset{\overset{O}{\|}}{C}-CH_2-CH_3$$

d. 3-Methylhexanal is a six-carbon aldehyde with a methyl group attached two carbons from the carbonyl group.

$$CH_3-CH_2-CH_2-\overset{\overset{CH_3}{|}}{CH}-CH_2-\overset{\overset{O}{\|}}{C}-H$$

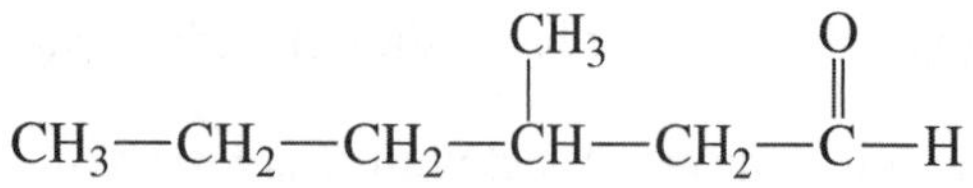

12.63 Compounds **a** and **b** are soluble in water because they each have a polar group with an oxygen atom that hydrogen bonds with water and fewer than five carbon atoms.

12.65 A chiral carbon is bonded to four different groups.

a. 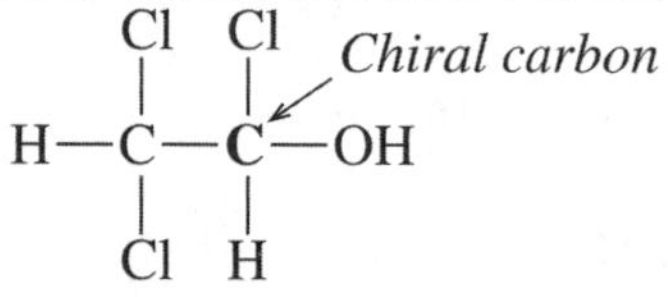 $$H-\overset{\overset{Cl}{|}}{\underset{\underset{Cl}{|}}{C}}-\overset{\overset{Cl}{|}}{\underset{\underset{H}{|}}{C}}-OH \quad \textit{Chiral carbon}$$

b. none
c. none

d. 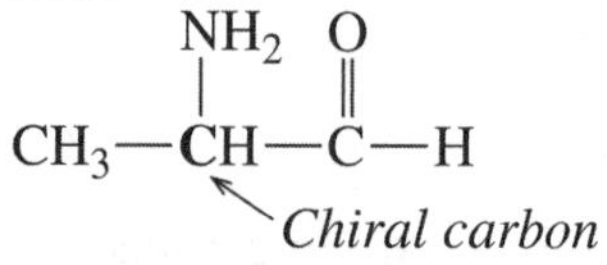 $$CH_3-\overset{\overset{NH_2}{|}}{CH}-\overset{\overset{O}{\|}}{C}-H \quad \textit{Chiral carbon}$$

e. 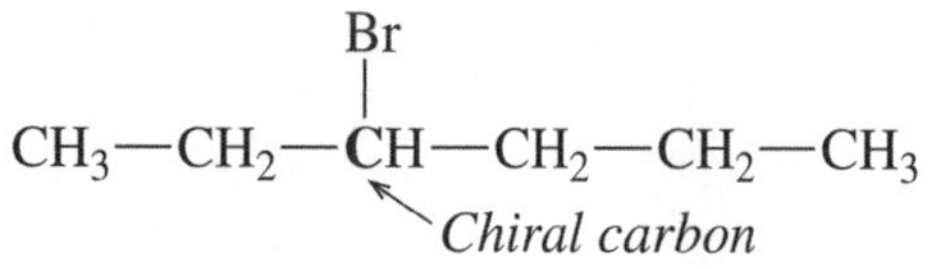 $$CH_3-CH_2-\overset{\overset{Br}{|}}{CH}-CH_2-CH_2-CH_3 \quad \textit{Chiral carbon}$$

12.67 Enantiomers are nonsuperimposable mirror images.
a. identical compounds (no chiral carbons)
b. enantiomers
c. enantiomers
d. identical compounds (no chiral carbons)

12.69 Primary alcohols oxidize to aldehydes and then to carboxylic acids. Secondary alcohols oxidize to ketones.

a. 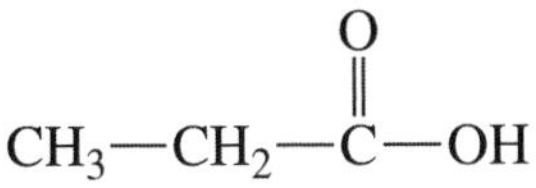 $CH_3-CH_2-\overset{\overset{\displaystyle O}{\|}}{C}-OH$

b. $CH_3-\overset{\overset{\displaystyle O}{\|}}{C}-CH_2-CH_2-CH_3$

c. $CH_3-CH_2-CH_2-\overset{\overset{\displaystyle O}{\|}}{C}-OH$

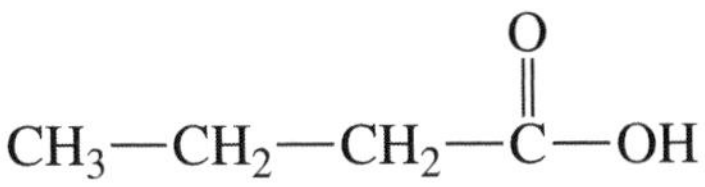

d.

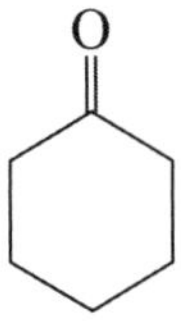

12.71

$CH_3-CH_2-CH_2-CH_2-CH_2-OH$ 1-Pentanol

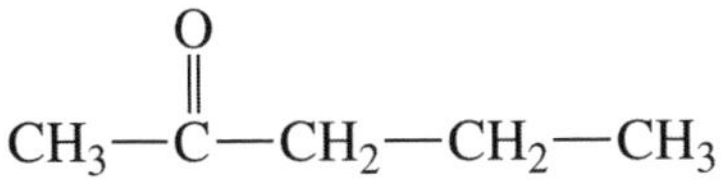

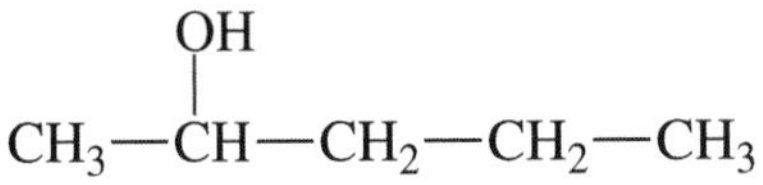

$CH_3-\overset{\overset{\displaystyle OH}{|}}{CH}-CH_2-CH_2-CH_3$ 2-Pentanol

$CH_3-CH_2-\overset{\overset{\displaystyle OH}{|}}{CH}-CH_2-CH_3$ 3-Pentanol

$HO-CH_2-\overset{\overset{\displaystyle CH_3}{|}}{CH}-CH_2-CH_3$ 2-Methyl-1-butanol

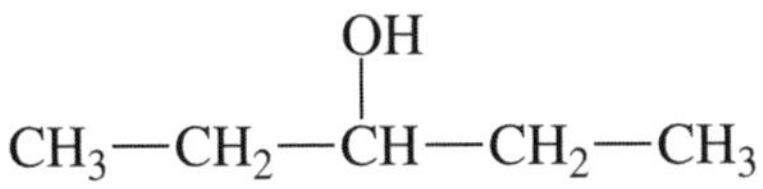

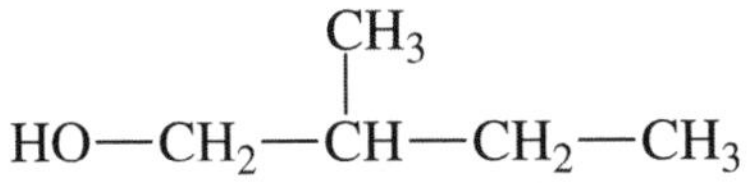

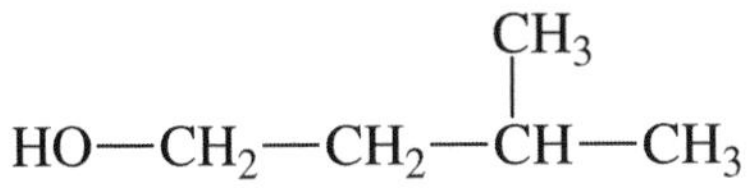

$HO-CH_2-CH_2-\overset{\overset{\displaystyle CH_3}{|}}{CH}-CH_3$ 3-Methyl-1-butanol

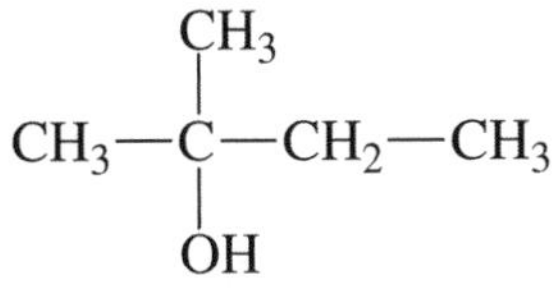

$CH_3-\underset{\underset{\displaystyle OH}{|}}{\overset{\overset{\displaystyle CH_3}{|}}{C}}-CH_2-CH_3$ 2-Methyl-2-butanol

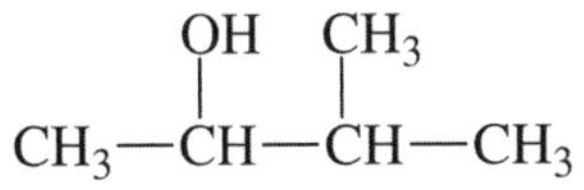

$CH_3-\overset{\overset{\displaystyle OH}{|}}{CH}-\overset{\overset{\displaystyle CH_3}{|}}{CH}-CH_3$ 3-Methyl-2-butanol

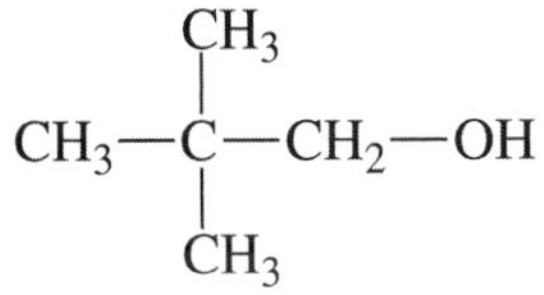

$CH_3-\underset{\underset{\displaystyle CH_3}{|}}{\overset{\overset{\displaystyle CH_3}{|}}{C}}-CH_2-OH$ 2,2-Dimethyl-1-propanol

113

12.73 Since the compound is synthesized from a primary alcohol and oxidizes to give a carboxylic acid, it must be an aldehyde.

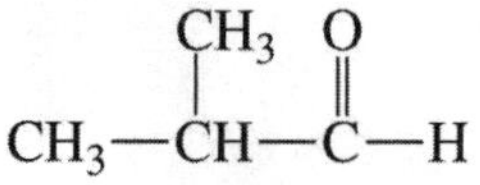

$$CH_3-\underset{\underset{\displaystyle CH_3}{|}}{CH}-\underset{\underset{\displaystyle O}{\|}}{C}-H$$ 2-Methylpropanal

12.75 **A** $CH_3-CH_2-CH_2-OH$ 1-Propanol

   B $CH_3-CH=CH_2$ Propene

   C $CH_3-CH_2-\underset{\underset{\displaystyle O}{\|}}{C}-H$ Propanal

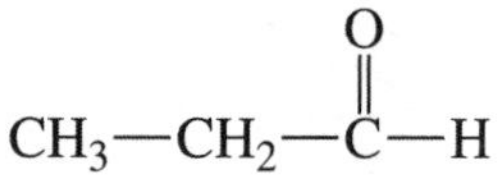

Answers to Combining Ideas from Chapters 9 to 12

CI.19 **a.**

Isotope	Number of Protons	Number of Neutrons	Number of Electrons
$^{27}_{14}\text{Si}$	14	13	14
$^{28}_{14}\text{Si}$	14	14	14
$^{29}_{14}\text{Si}$	14	15	14
$^{30}_{14}\text{Si}$	14	16	14
$^{31}_{14}\text{Si}$	14	17	14

   b. Electron arrangement of Si: 2,8,4

   c. $^{28}_{14}\text{Si}$ $27.98 \times \dfrac{92.23}{100} = 25.81$ amu

   $^{29}_{14}\text{Si}$ $28.98 \times \dfrac{4.67}{100} = 1.35$ amu

   $^{30}_{14}\text{Si}$ $29.97 \times \dfrac{3.10}{100} = 0.929$ amu

   Atomic mass of Si $= \overline{28.09 \text{ amu}}$

   d. $^{27}_{14}\text{Si} \rightarrow \, ^{27}_{13}\text{Al} + \, ^{0}_{+1}e$

   $^{31}_{14}\text{Si} \rightarrow \, ^{31}_{15}\text{P} + \, ^{0}_{-1}e$

   e.

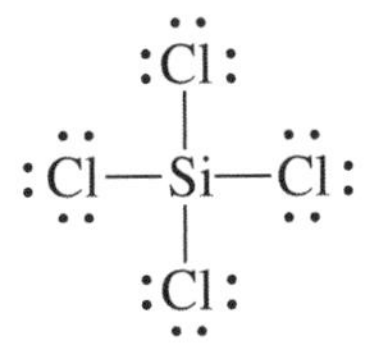

$$\ddot{:}\overset{\displaystyle \cdots}{\text{Cl}}\ddot{:}$$

$$:\ddot{\text{Cl}}-\overset{\displaystyle |}{\underset{\displaystyle |}{\text{Si}}}-\ddot{\text{Cl}}:$$

$$:\ddot{\text{Cl}}:$$

   With 4 bonding pairs and no lone pairs, the shape of SiCl_4 is tetrahedral.

   f. Half of a radioactive sample decays with each half-life:

   16 μCi of $^{31}_{14}\text{Si}$ $\xrightarrow{\text{1 half-life}}$ 8.0 μCi of $^{31}_{14}\text{Si}$ $\xrightarrow{\text{1 half-life}}$ 4.0 μCi of $^{31}_{14}\text{Si}$ $\xrightarrow{\text{1 half-life}}$

   2.0 μCi of $^{31}_{14}\text{Si}$

   Therefore, 3 half-lives have passed.

   $3 \, \cancel{\text{half-lives}} \times \dfrac{2.6 \text{ h}}{1 \, \cancel{\text{half-life}}} = 7.8$ h (2 SFs)

CI.21 **a.** $^{226}_{88}\text{Ra} \rightarrow \, ^{222}_{86}\text{Rn} + \, ^{4}_{2}\text{He}$

   b. $^{222}_{86}\text{Rn} \rightarrow \, ^{218}_{84}\text{Po} + \, ^{4}_{2}\text{He}$

   c. First determine how many half-lives have passed:

   $15.2 \, \cancel{\text{days}} \times \dfrac{1 \text{ half-life}}{3.8 \, \cancel{\text{days}}} = 4.0$ half-lives

   The number of atoms of radon-222 that remain:

$$24\,000 \text{ atoms} \times \tfrac{1}{2} \times \tfrac{1}{2} \times \tfrac{1}{2} \times \tfrac{1}{2} = 24\,000 \text{ atoms} \times \tfrac{1}{16} = 1500 \text{ atoms of radon-222 (2 SFs)}$$

d. Volume of room = 72 000 L

$$72\,000 \text{ L air} \times \frac{2.5 \text{ pCi}}{1 \text{ L air}} \times \frac{1 \text{ Ci}}{10^{12} \text{ pCi}} \times \frac{3.7 \times 10^{10} \text{ disintegrations}/s}{1 \text{ Ci}}$$

$$\times \frac{1 \text{ alpha particle}}{1 \text{ disintegration}} \times \frac{60 \text{ s}}{1 \text{ min}} \times \frac{60 \text{ min}}{1 \text{ h}} \times \frac{24 \text{ h}}{1 \text{ day}}$$

$$= 5.8 \times 10^{8} \text{ alpha particles per day (2 SFs)}$$

CI.23 **a.** $CH_3\!-\!CH_2\!-\!CH_2\!-\!\overset{\displaystyle O}{\overset{\|}{C}}\!-\!H$

b.

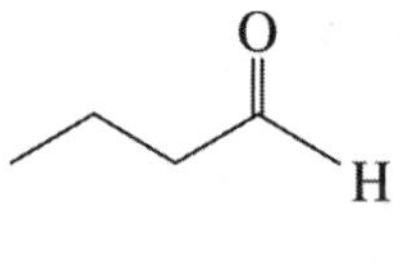

c. butanal

d. $CH_3\!-\!CH_2\!-\!CH_2\!-\!CH_2\!-\!OH$

13

Carbohydrates

13.1 Photosynthesis requires CO_2, H_2O, and the energy from the Sun. Respiration requires O_2 from the air and glucose from our foods.

13.3 Monosaccharides are composed of a chain of three to eight carbon atoms, one in a carbonyl group as an aldehyde or ketone, and the rest attached to hydroxyl groups. A monosaccharide cannot be split or hydrolyzed into smaller carbohydrates. A disaccharide consists of two monosaccharide units joined together. A disaccharide can be hydrolyzed into 2 monosaccharide units.

13.5 Hydroxyl groups are found in all monosaccharides along with a carbonyl group on the first or second carbon.

13.7 A ketopentose has five carbon atoms, several hydroxyl groups, and a ketone group.

13.9 **a.** This six-carbon monosaccharide has a carbonyl group on carbon 2; it is a ketohexose.
 b. This five-carbon monosaccharide has a carbonyl group on carbon 1; it is an aldopentose.

13.11 A Fischer projection is a two-dimensional representation of the three-dimensional structure of a molecule. In the D isomer, the —OH on the chiral carbon atom at the bottom of the chain is on the right side, whereas in the L isomer, the —OH appears on the left side.

13.13 **a.** This structure is a D isomer since the hydroxyl group on the chiral carbon farthest from the carbonyl group is on the right.
 b. This structure is a D isomer since the hydroxyl group on the chiral carbon farthest from the carbonyl group is on the right.
 c. This structure is an L isomer since the hydroxyl group on the chiral carbon farthest from the carbonyl group is on the left.
 d. This structure is a D isomer since the hydroxyl group on the chiral carbon farthest from the carbonyl group is on the right.

13.15 **a.**

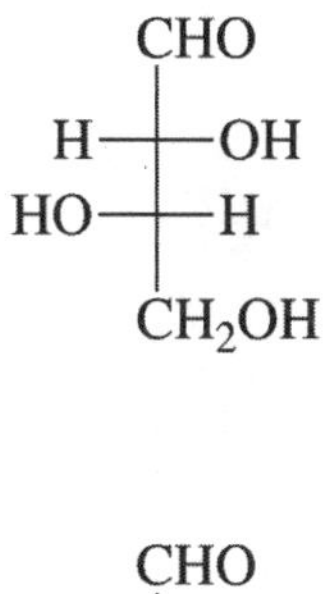

```
        CHO
    H ——|—— OH
   HO ——|—— H
        CH2OH
```

b.

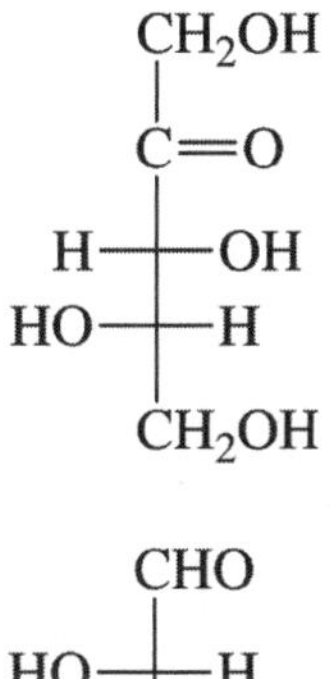

```
        CH2OH
         |
        C=O
    H ——|—— OH
   HO ——|—— H
        CH2OH
```

c.

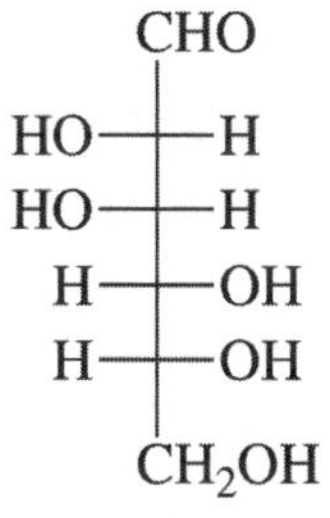

```
        CHO
   HO ——|—— H
   HO ——|—— H
    H ——|—— OH
    H ——|—— OH
        CH2OH
```

d.

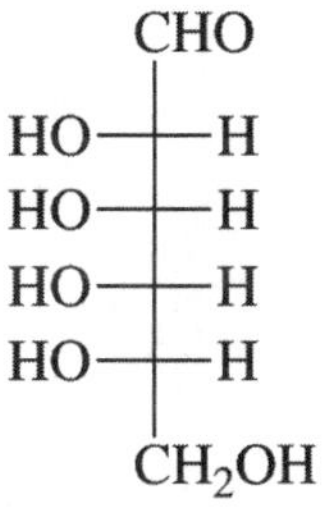

```
        CHO
   HO ——|—— H
   HO ——|—— H
   HO ——|—— H
   HO ——|—— H
        CH2OH
```

13.17 L-glucose is the mirror image of D-glucose.

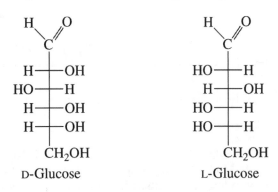

13.19 In D-galactose, the hydroxyl group on carbon 4 extends to the left; in D-glucose, this hydroxyl group goes to the right.

13.21 **a.** Glucose is also called blood sugar.
b. Galactose is not metabolized in the condition called galactosemia.
c. Fructose is also called fruit sugar.

13.23 In the cyclic structure of glucose, there are five carbon atoms and an oxygen atom.

13.25 In the α isomer, the hydroxyl ($-OH$) on carbon 1 is drawn down; in the β isomer, the hydroxyl ($-OH$) on carbon 1 is drawn up.

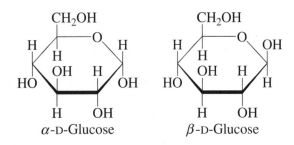

13.27 **a.** This is the α isomer because the $-OH$ on carbon 2 is down.
b. This is the α isomer because the $-OH$ on carbon 1 is down.

13.29

$$
\begin{array}{c}
CH_2OH \\
| \\
H-C-OH \\
| \\
HO-C-H \qquad \text{D-Xylitol}\\
| \\
H-C-OH \\
| \\
CH_2OH
\end{array}
$$

13.31 Oxidation product: Reduction product (sugar alcohol):

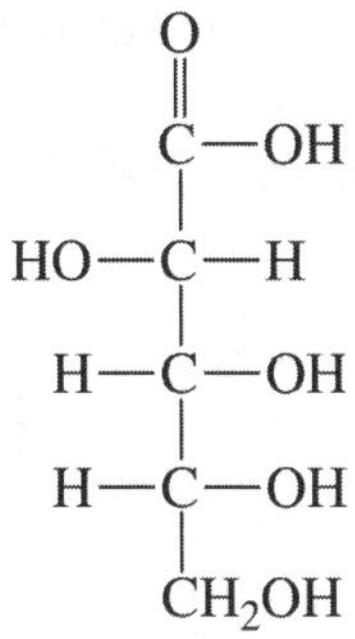

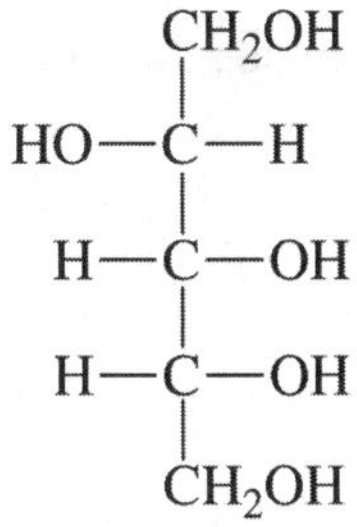

$$
\begin{array}{c}
\overset{\displaystyle O}{\underset{\displaystyle \parallel}{}}\\
\text{C—OH}\\
\text{HO—C—H}\\
\text{H—C—OH}\\
\text{H—C—OH}\\
\text{CH}_2\text{OH}
\end{array}
\qquad\qquad
\begin{array}{c}
\text{CH}_2\text{OH}\\
\text{HO—C—H}\\
\text{H—C—OH}\\
\text{H—C—OH}\\
\text{CH}_2\text{OH}
\end{array}
\qquad \text{D-Arabitol}
$$

13.33 **a.** When this disaccharide is hydrolyzed, galactose and glucose are produced. The glycosidic bond is a β-1,4 bond since the ether bond is drawn up from carbon 1 of the galactose unit, which is on the left in the drawing, to carbon 4 of the glucose on the right. β-Lactose is the name of this disaccharide, since the hydroxyl group on carbon 1 of the glucose unit is drawn up.

      b. When this disaccharide is hydrolyzed, two molecules of glucose are produced. The glycosidic bond is an α-1,4 bond since the ether bond is drawn down from carbon 1 of the glucose unit on the left to carbon 4 of the glucose on the right. α-Maltose is the name of this disaccharide, since the hydroxyl group on the rightmost glucose unit is drawn down.

13.35 **a.** β-Lactose is a reducing sugar; the ring on the right can open up to form an aldehyde that can undergo oxidation.

      b. α-Maltose is a reducing sugar; the ring on the right can open up to form an aldehyde that can undergo oxidation.

13.37 **a.** Another name for table sugar is sucrose.

      b. Lactose is the disaccharide found in milk and milk products.

      c. Maltose is also called malt sugar.

      d. When lactose is hydrolyzed, the products are the monosaccharides galactose and glucose.

13.39 **a.** Amylose is an unbranched polymer of glucose units joined by α-1,4-glycosidic bonds. Amylopectin is a branched polymer of glucose joined by α-1,4 and α-1,6-glycosidic bonds.

      b. Amylopectin, which is produced in plants, is a branched polymer of glucose, joined by α-1,4-glycosidic and α-1,6-glycosidic bonds. Glycogen, which is produced in animals, is a highly branched polymer of glucose, joined by α-1,4- and α-1,6-glycosidic bonds.

13.41 **a.** Cellulose is not digestible by humans since we do not have the enzymes necessary to break the β-1,4-glycosidic bonds in cellulose.

      b. Amylose and amylopectin are the storage forms of carbohydrates in plants.

      c. Amylose is the polysaccharide that contains only α-1,4-glycosidic bonds.

      d. Glycogen contains many α-1,4- and α-1,6-glycosidic bonds and is the most highly branched.

13.43 **a.** Isomaltose is a disaccharide.

      b. Isomaltose consists of two α-D-glucose units.

      c. The glycosidic link in isomaltose is an α-1,6-glycosidic bond.

      d. The structure shown is α-isomaltose.

      e. α-Isomaltose is a reducing sugar; the ring on the right can open up to form an aldehyde that can undergo oxidation.

13.45   **a.**   Melezitose is a trisaccharide.

   b.   Melezitose contains two units of the aldohexose α-D-glucose and one unit of the ketohexose β-D-fructose.

   c.   Melezitose, like sucrose, is not a reducing sugar; the rings on the right can not open up to form an aldehyde that can undergo oxidation.

13.47   D-Fructose is a ketohexose with the carbonyl group on carbon 2; D-galactose is an aldohexose where the carbonyl group is on carbon 1. In the Fischer projection of D-galactose, the —OH group on carbon 4 is drawn on the left; in fructose, the —OH is on the right.

13.49   D-Galactose is the mirror image of L-galactose. In the Fischer projection of D-galactose, the —OH groups on carbon 2 and carbon 5 are drawn on the right side, but are on the left for carbon 3 and carbon 4. In L-galactose, the —OH groups are reversed: carbon 2 and carbon 5 have —OH groups on the left, and carbon 3 and carbon 4 have —OH groups on the right.

13.51   **a.**

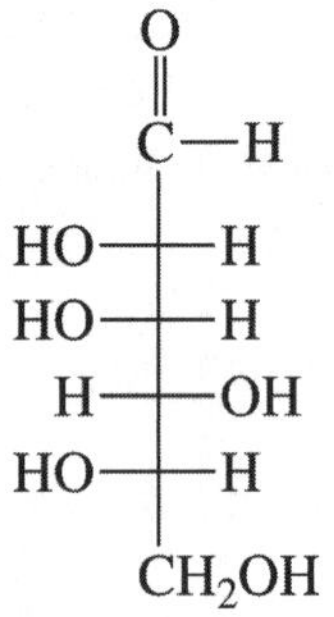

```
            O
            ‖
            C—H
   HO———H
   HO———H
    H———OH
   HO———H
        CH2OH
      L-Gulose
```

   b.

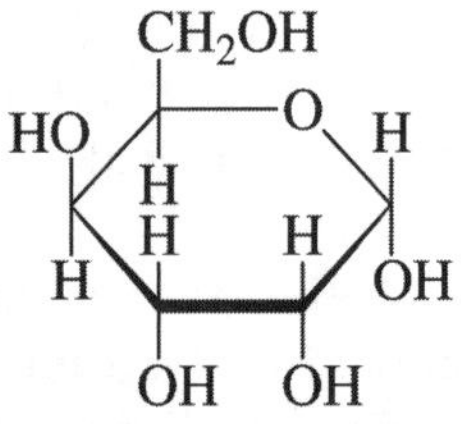

```
      CH2OH                    CH2OH
   HO      O  H             HO      O  OH
      H                        H
      H    H                   H    H
    H        OH              H        H
      OH  OH                   OH  OH
   α-D-Gulose                β-D-Gulose
```

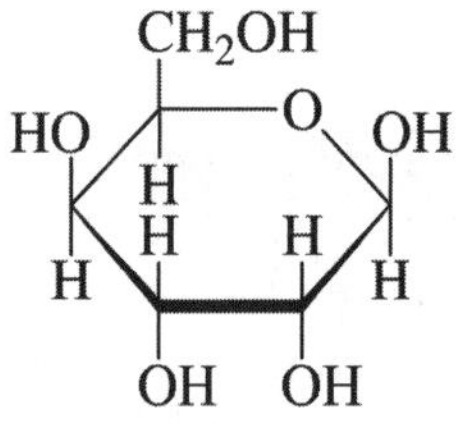

13.53   Since D-sorbitol can be oxidized to D-glucose, it must contain the same number of carbons with the same groups attached as glucose. The difference is that sorbitol has only hydroxyl groups while glucose has an aldehyde group. In sorbitol, the aldehyde group is changed to a hydroxyl.

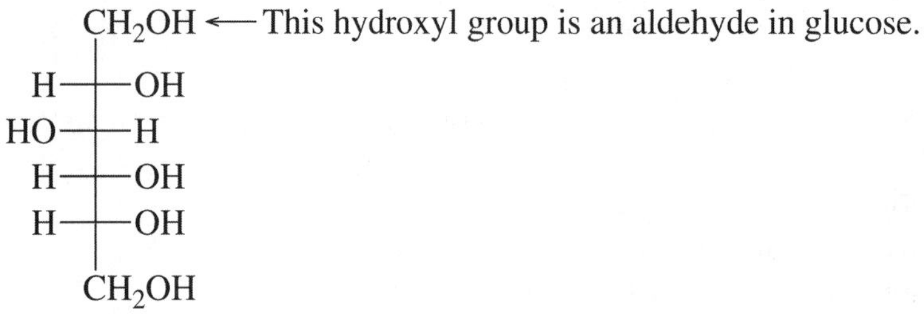

```
      CH2OH ←— This hydroxyl group is an aldehyde in glucose.
    H———OH
   HO———H
    H———OH
    H———OH
      CH2OH
```

13.55   When the α-galactose forms an open chain structure, it can close to form either α- or β-galactose.

13.57

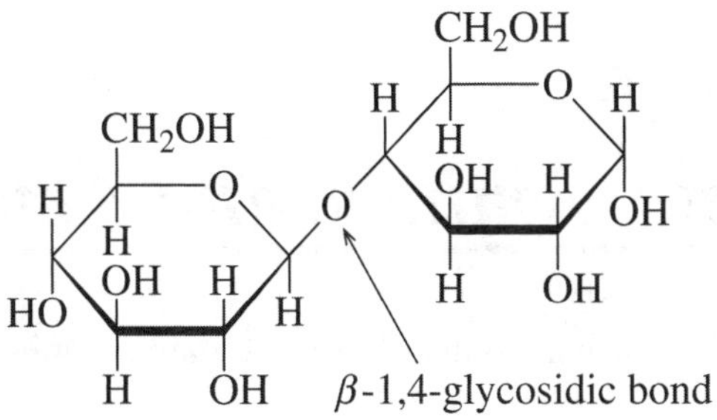

α-Cellobiose

β-1,4-glycosidic bond

13.59 a.

β-1,6-glycosidic bond

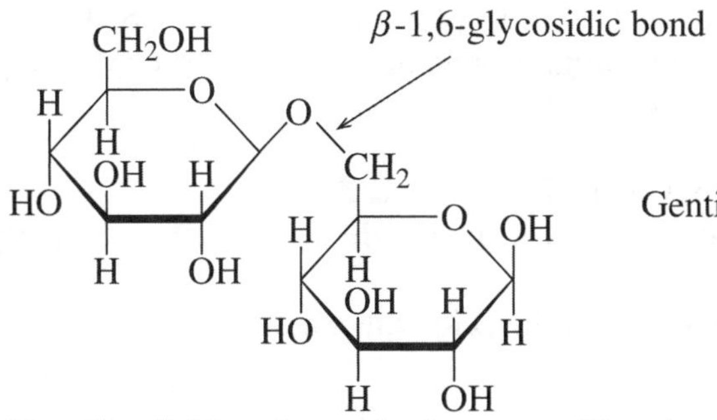

Gentiobiose

b. Yes. Gentiobiose is a reducing sugar. The ring on the right can open up to form an aldehyde that can undergo oxidation.

Carboxylic Acids, Esters, Amines, and Amides

14.1 Methanoic acid (formic acid) is the carboxylic acid that is responsible for the pain associated with ant stings.

14.3 Each compound contains three carbon atoms. They differ because propanal, an aldehyde, contains a carbonyl group bonded to a hydrogen; in propanoic acid, the carbonyl group connects to a hydroxyl group, forming a carboxyl group.

14.5 **a.** Ethanoic acid (acetic acid) is the carboxylic acid with two carbons.
    b. 2,4-Dibromobenzoic acid is an aromatic carboxylic acid with bromine atoms on carbon 2 and carbon 4 of the ring.
    c. 4-Methylpentanoic acid is a five-carbon carboxylic acid with a methyl group on carbon 4 of the chain.

14.7 **a.** Benzoic acid is the carboxylic acid of benzene.

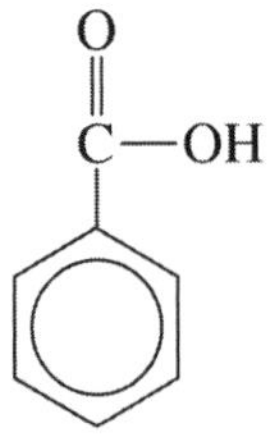

$$\underset{\overset{||}{\underset{\text{O}}{}}}{\text{C}}\text{—OH}$$

    b. Chloroethanoic acid is a carboxylic acid that has a two-carbon chain with a chlorine atom on carbon 2.

$$\text{Cl—CH}_2\text{—}\overset{\overset{\text{O}}{||}}{\text{C}}\text{—OH}$$

    c. Butyric acid has four carbons.

$$\text{CH}_3\text{—CH}_2\text{—CH}_2\text{—}\overset{\overset{\text{O}}{||}}{\text{C}}\text{—OH}$$

    d. Heptanoic acid is a carboxylic acid that has a seven-carbon chain.

$$\text{CH}_3\text{—CH}_2\text{—CH}_2\text{—CH}_2\text{—CH}_2\text{—CH}_2\text{—}\overset{\overset{\text{O}}{||}}{\text{C}}\text{—OH}$$

14.9 **a.** Propanoic acid is the most soluble of the group because it has the fewest number of carbon atoms in its hydrocarbon chain. Solubility of carboxylic acids decreases as the number of carbon atoms in the hydrocarbon chain increases.
    b. Propanoic acid is more soluble than 1-hexanol because it has fewer carbon atoms in its hydrocarbon chain. Propanoic acid is also more soluble because the carboxyl group forms more hydrogen bonds with water than does the hydroxyl group of an alcohol. An alkane is not soluble in water.

14.11 **a.** $CH_3-CH_2-CH_2-CH_2-\overset{\overset{\displaystyle O}{\|}}{C}-OH + H_2O \rightleftarrows CH_3-CH_2-CH_2-CH_2-\overset{\overset{\displaystyle O}{\|}}{C}-O^- + H_3O^+$

b. $CH_3-\overset{\overset{\displaystyle O}{\|}}{C}-OH + H_2O \rightleftarrows CH_3-\overset{\overset{\displaystyle O}{\|}}{C}-O^- + H_3O^+$

14.13 **a.** $H-\overset{\overset{\displaystyle O}{\|}}{C}-OH + NaOH \longrightarrow H-\overset{\overset{\displaystyle O}{\|}}{C}-O^-Na^+ + H_2O$

b. $Cl-CH_2-CH_2-\overset{\overset{\displaystyle O}{\|}}{C}-OH + NaOH \longrightarrow Cl-CH_2-CH_2-\overset{\overset{\displaystyle O}{\|}}{C}-O^-Na^+ + H_2O$

c. (benzene ring)$-\overset{\overset{\displaystyle O}{\|}}{C}-OH$ + NaOH $\longrightarrow$ (benzene ring)$-\overset{\overset{\displaystyle O}{\|}}{C}-O^-Na^+$ + H_2O

14.15 A carboxylic acid reacts with an alcohol to form an ester and water. In an ester, the —H of the carboxylic acid is replaced by an alkyl group.

a. $CH_3-\overset{\overset{\displaystyle O}{\|}}{C}-O-CH_3$

b. $CH_3-CH_2-CH_2-\overset{\overset{\displaystyle O}{\|}}{C}-O-CH_3$

c. (benzene ring)$-\overset{\overset{\displaystyle O}{\|}}{C}-O-CH_3$

14.17 A carboxylic acid and an alcohol react to give an ester with the elimination of water.

a. $CH_3-CH_2-CH_2-CH_2-\overset{\overset{\displaystyle O}{\|}}{C}-O-\overset{\overset{\displaystyle CH_3}{|}}{CH}-CH_3$

b. $CH_3-CH_2-\overset{\overset{\displaystyle O}{\|}}{C}-O-CH_2-CH_2-CH_3$

14.19 **a.** The alcohol part of the ester is from methanol (methyl alcohol) and the carboxylic acid part is from methanoic acid (formic acid). The ester is named methyl methanoate (methyl formate).
b. The alcohol part of the ester is from methanol (methyl alcohol) and the carboxylic acid part is from ethanoic acid (acetic acid). The ester is named methyl ethanoate (methyl acetate).
c. The alcohol part of the ester is from propanol (propyl alcohol) and the carboxylic acid part is from ethanoic acid (acetic acid). The ester is named propyl ethanoate (propyl acetate).

14.21 **a.** The alkyl part of the ester comes from the three-carbon 1-propanol and the carboxylate part from the four-carbon butyric acid.

$$CH_3—CH_2—CH_2—\overset{\overset{\displaystyle O}{\|}}{C}—O—CH_2—CH_2—CH_3$$

b. The alkyl part of the ester comes from the four-carbon 1-butanol and the carboxylate part from the one-carbon formic acid.

$$H—\overset{\overset{\displaystyle O}{\|}}{C}—O—CH_2—CH_2—CH_2—CH_3$$

c. The alkyl part of the ester comes from the two-carbon ethanol and the carboxylate part from the five-carbon pentanoic acid.

$$CH_3—CH_2—CH_2—CH_2—\overset{\overset{\displaystyle O}{\|}}{C}—O—CH_2—CH_3$$

14.23 **a.** The flavor and odor of bananas is pentyl ethanoate (pentyl acetate).
   b. The flavor and odor of oranges is octyl ethanoate (octyl acetate).
   c. The flavor and odor of pears is propyl ethanoate (propyl acetate).

14.25 Acid hydrolysis of an ester adds water in the presence of acid and gives a carboxylic acid and an alcohol.

14.27 Acid hydrolysis of an ester gives the carboxylic acid and the alcohol which were combined to form the ester; basic hydrolysis of an ester gives the salt of carboxylic acid and the alcohol which combine to form the ester.

a. $CH_3—CH_2—\overset{\overset{\displaystyle O}{\|}}{C}—O^-Na^+$ and $CH_3—OH$

b. $CH_3—\overset{\overset{\displaystyle O}{\|}}{C}—OH$ and $CH_3—CH_2—CH_2—OH$

c. $CH_3—CH_2—CH_2—\overset{\overset{\displaystyle O}{\|}}{C}—OH$ and $CH_3—CH_2—OH$

d.

and $CH_3—CH_2—OH$

14.29 **a.** This is a primary (1°) amine; there is only one alkyl group attached to the nitrogen atom.
   b. This is a secondary (2°) amine; there are two alkyl groups attached to the nitrogen atom.
   c. This is a tertiary (3°) amine; three are two alkyl groups and one aromatic group attached to the nitrogen atom.
   d. This is a tertiary (3°) amine; there are three alkyl groups attached to the nitrogen atom.

14.31 The common name of an amine consists of naming the alkyl groups bonding to the nitrogen atom in alphabetical order.
   a. A three-carbon alkyl group attached to $—NH_2$ is propylamine.
   b. A one-carbon and a three-carbon alkyl group attached to nitrogen form methylpropylamine.
   c. A one-carbon and two two-carbon alkyl groups attached to nitrogen form diethylmethylamine.

14.33 **a.** $CH_3-CH_2-NH_2$

b.

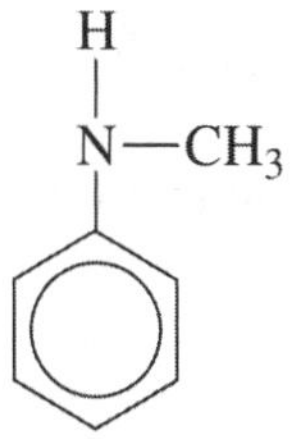

$$N-CH_3$$ (with H above N, attached to benzene ring)

c. $CH_3-CH_2-CH_2-CH_2-\overset{H}{\underset{|}{N}}-CH_2-CH_2-CH_3$

14.35 **a.** Yes; amines with fewer than six carbon atoms hydrogen bond with water molecules and are soluble in water.
   b. Yes; amines with fewer than six carbon atoms hydrogen bond with water molecules and are soluble in water.
   c. No; an amine with eight carbon atoms has large hydrocarbon sections that make it insoluble in water.
   d. Yes; amines with fewer than six carbon atoms hydrogen bond with water molecules and are soluble in water.

14.37 Amines, which are weak bases, bond with a proton from water to give a hydroxide ion and an ammonium ion.
   a. $CH_3-NH_2 + H_2O \rightleftharpoons CH_3-NH_3^+ + OH^-$
   b. $CH_3-NH-CH_3 + H_2O \rightleftharpoons CH_3-\overset{+}{N}H_2-CH_3 + OH^-$
   c. (benzene ring with NH_2) $+ H_2O \rightleftharpoons$ (benzene ring with NH_3^+) $+ OH^-$

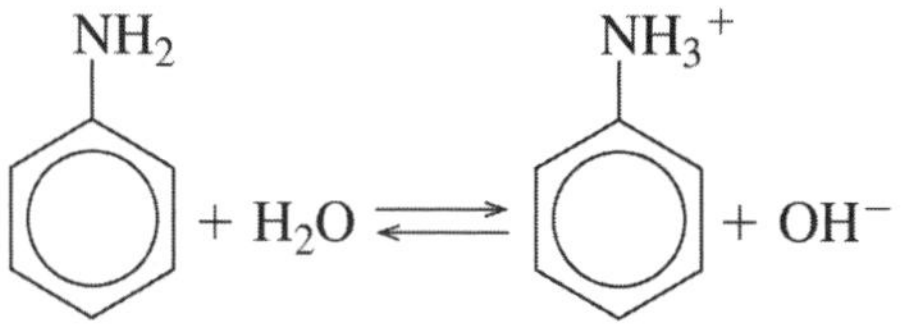

14.39 The heterocyclic amines contain one or more nitrogen atoms in a ring.
   a. Pyrimidine is a six-atom ring with two nitrogen atoms and three double bonds.
   b. Anabasine contains both a pyridine ring (one N and three double bonds in a ring of six atoms) and a piperidine ring (one N in a saturated ring of six atoms).

14.41 The five-atom ring with one nitrogen atom and two double bonds is pyrrole.

14.43 Carboxylic acids react with amines to form amides with the elimination of water.
   a. $CH_3-\overset{O}{\overset{||}{C}}-NH_2$
   b. $CH_3-\overset{O}{\overset{||}{C}}-\overset{H}{\underset{|}{N}}-CH_2-CH_3$
   c. (benzene ring)$-\overset{O}{\overset{||}{C}}-\overset{H}{\underset{|}{N}}-CH_2-CH_2-CH_3$

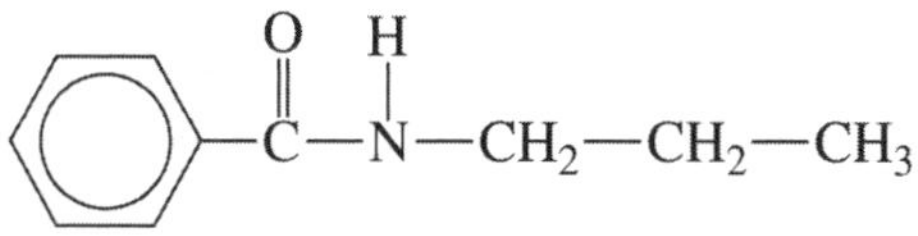

14.45 **a.** Ethanamide (acetamide) is a chain of two carbon atoms with the carbonyl carbon bonded to an amino group.

      b. Butanamide (butyramide) is a chain of four carbon atoms with the carbonyl carbon bonded to an amino group.

      c. Methanamide (formamide) has only the carbonyl carbon bonded to an amino group.

14.47 **a.** This is an amide of propionic acid, which has three carbon atoms.

$$CH_3-CH_2-\overset{\overset{\displaystyle O}{\|}}{C}-NH_2$$

      b. This is an amide of pentanoic acid, which has five carbon atoms, with a methyl group attached to carbon 2 of the chain (counting from the carbonyl carbon).

$$CH_3-CH_2-CH_2-\overset{\overset{\displaystyle CH_3}{|}}{CH}-\overset{\overset{\displaystyle O}{\|}}{C}-NH_2$$

      c. This is the simplest of the amides with only one carbon atom.

$$H-\overset{\overset{\displaystyle O}{\|}}{C}-NH_2$$

14.49 Acid hydrolysis of an amide gives the carboxylic acid and the amine salt.

      a. $CH_3-\overset{\overset{\displaystyle O}{\|}}{C}-OH + NH_4{}^+Cl^-$

      b. $CH_3-CH_2-\overset{\overset{\displaystyle O}{\|}}{C}-OH + NH_4{}^+Cl^-$

      c. $CH_3-CH_2-CH_2-\overset{\overset{\displaystyle O}{\|}}{C}-OH + CH_3-NH_3{}^+Cl^-$

      d. 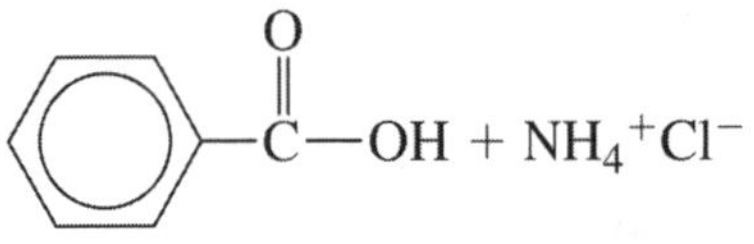 $\bigcirc\!\!-\overset{\overset{\displaystyle O}{\|}}{C}-OH + NH_4{}^+Cl^-$

      e. $CH_3-CH_2-CH_2-CH_2-\overset{\overset{\displaystyle O}{\|}}{C}-OH + CH_3-CH_2-NH_3{}^+Cl^-$

14.51 $CH_3-CH_2-CH_2-NH_2$ $CH_3-CH_2-NH-CH_3$ $CH_3-\overset{\overset{\displaystyle CH_3}{|}}{N}-CH_3$

       Propylamine (1°)   Ethylmethylamine (2°)   Trimethylamine (3°)

      $CH_3-\overset{\overset{\displaystyle CH_3}{|}}{CH}-NH_2$ Isopropylamine (1°)

14.53 aromatic, alcohol, amine

14.55 **a.** 3-methylbutanoic acid; 3-methylbutyric acid
      b. ethyl benzoate
      c. ethyl propanoate; ethyl propionate
      d. 2-chlorobenzoic acid
      e. pentanoic acid

14.57 **a.** $CH_3-\overset{\overset{O}{\|}}{C}-O-CH_3$

b. $CH_3-CH_2-CH_2-\overset{\overset{O}{\|}}{C}-O-CH_2-CH_3$

c. $CH_3-CH_2-\overset{\overset{CH_3}{|}}{CH}-CH_2-\overset{\overset{O}{\|}}{C}-OH$

d. 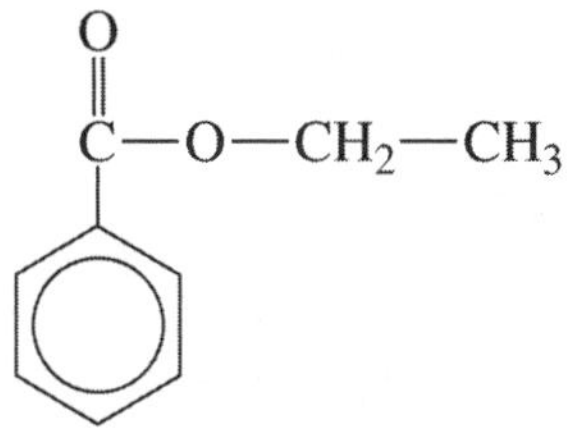 $\overset{\overset{O}{\|}}{C}-O-CH_2-CH_3$ (attached to benzene ring)

14.59 **a.** $CH_3-CH_2-\overset{\overset{O}{\|}}{C}-O^-K^+ + H_2O$

b. $CH_3-CH_2-\overset{\overset{O}{\|}}{C}-O-CH_3 + H_2O$

c. 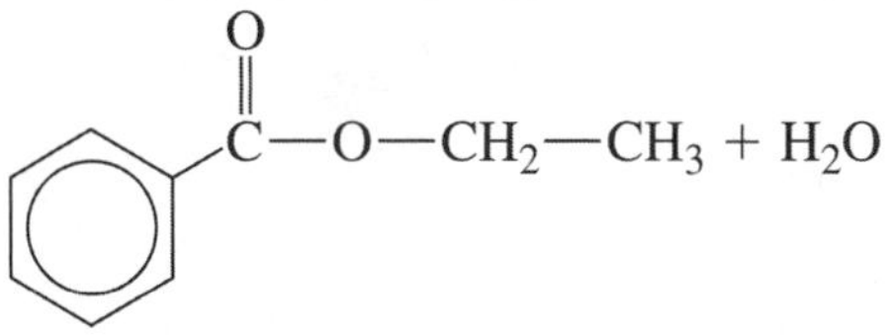 $\overset{\overset{O}{\|}}{C}-O-CH_2-CH_3 + H_2O$ (attached to benzene ring)

d. $CH_3-CH_2-\overset{\overset{O}{\|}}{C}-\overset{\overset{H}{|}}{N}-CH_3 + H_2O$

14.61 **a.** $CH_3-CH_2-\overset{\overset{O}{\|}}{C}-OH + HO-\overset{\overset{CH_3}{|}}{CH}-CH_3$

b. $CH_3-\overset{\overset{CH_3}{|}}{CH}-\overset{\overset{O}{\|}}{C}-OH + HO-CH_2-CH_2-CH_3$

c. 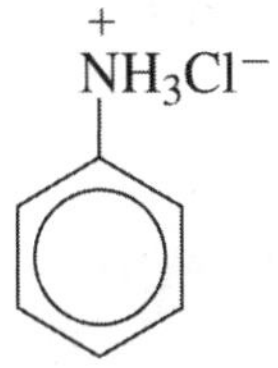 $\overset{+}{N}H_3Cl^-$ (attached to benzene ring)

14.63 **a.** Two methyl groups are bonded to a nitrogen atom.

$CH_3-\overset{\overset{H}{|}}{N}-CH_3$

b. The amino group is attached to a six-carbon cycloalkane ring.

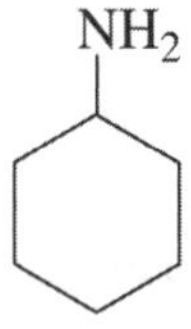

NH_2 (attached to cyclohexane ring)

127

c. This is an ammonium salt with two methyl groups bonded to the nitrogen atom.

$$CH_3—\overset{\displaystyle CH_3}{\overset{|}{N}H_2}{}^+Cl^-$$

d. Three ethyl groups are bonded to a nitrogen atom.

$$CH_3—CH_2—\overset{\displaystyle CH_2—CH_3}{\overset{|}{N}}—CH_2—CH_3$$

e. An ethyl group is bonded to the nitrogen atom of aniline.

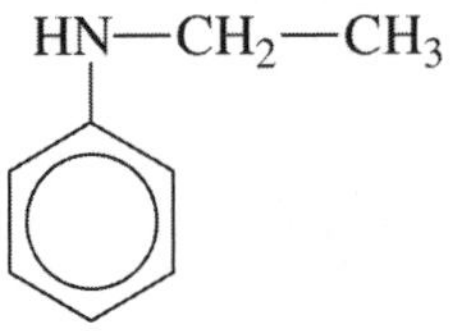

$$HN—CH_2—CH_3$$

14.65 a. An amine accepts a proton from water, which produces an ammonium ion and OH^-.

$$CH_3—CH_2—NH_3{}^+ + OH^-$$

b. The amine accepts a proton to give an ammonium salt.

$$CH_3—CH_2—NH_3{}^+Cl^-$$

c. An amine accepts a proton from water, which produces an ammonium ion and OH^-.

$$CH_3—CH_2—\overset{+}{N}H_2—CH_3 + OH^-$$

d. The amine accepts a proton to give an ammonium salt.

$$CH_3—CH_2—\overset{+}{N}H_2—CH_3Cl^-$$

14.67 aromatic, amine, carboxylate salt

14.69 a. $CH_3—\overset{\displaystyle O}{\overset{||}{C}}—O—CH_2—CH_2—CH_3$

b. $CH_3—\overset{\displaystyle O}{\overset{||}{C}}—OH + HO—CH_2—CH_2—CH_3 \underset{Heat}{\overset{H^+}{\rightleftarrows}} CH_3—\overset{\displaystyle O}{\overset{||}{C}}—O—CH_2—CH_2—CH_3 + H_2O$

c. $CH_3—\overset{\displaystyle O}{\overset{||}{C}}—O—CH_2—CH_2—CH_3 + H_2O \overset{H^+}{\rightleftarrows} CH_3—\overset{\displaystyle O}{\overset{||}{C}}—OH + HO—CH_2—CH_2—CH_3$

d. $CH_3—\overset{\displaystyle O}{\overset{||}{C}}—O—CH_2—CH_2—CH_3 + NaOH \overset{Heat}{\longrightarrow} CH_3—\overset{\displaystyle O}{\overset{||}{C}}—O^-Na^+$
$$+ HO—CH_2—CH_2—CH_3$$

e. Molar mass of propyl acetate ($C_5H_{10}O_2$) = 5(12.0 g) + 10(1.01 g) + 2(16.0 g) = 102.1 g/mole

$$1.58 \text{ g } C_5H_{10}O_2 \times \frac{1 \text{ mole } C_5H_{10}O_2}{102.1 \text{ g } C_5H_{10}O_2} \times \frac{1 \text{ mole NaOH}}{1 \text{ mole } C_5H_{10}O_2} \times \frac{1000 \text{ mL solution}}{0.208 \text{ mole NaOH}}$$

= 74.4 mL of a 0.208 M NaOH solution (3 SFs)

14.71 a. $H_2N—\overset{\displaystyle O}{\underset{}{\bigcirc}}—\overset{\displaystyle O}{\overset{||}{C}}—O—CH_2—CH_2—\overset{+}{N}\overset{\diagup CH_2—CH_3}{\underset{\diagdown CH_2—CH_3}{—H}}Cl^-$

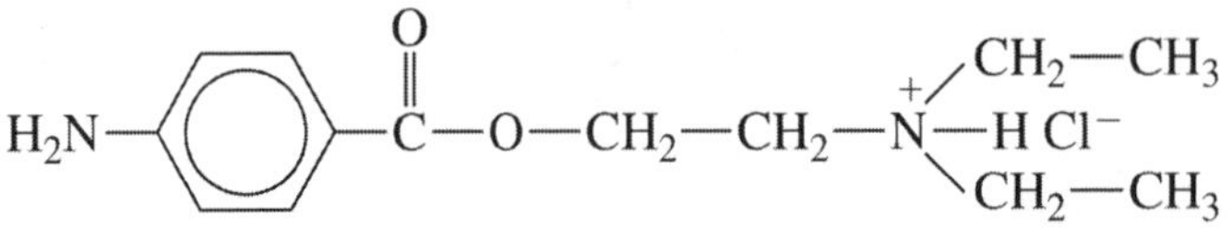

b. The amine salt (Novocaine) is more soluble in water and body fluids than the amine procaine.

15

Lipids

15.1 Lipids provide energy, and protection and insulation for the organs in the body. Lipids are also an important component of cell membranes.

15.3 Because lipids are not soluble in water, a polar solvent, they are nonpolar molecules.

15.5 All fatty acids contain a long chain of carbon atoms with a carboxylic acid group. Saturated fatty acids contain only carbon–carbon single bonds; unsaturated fatty acids contain one or more double bonds.

15.7 **a.** Palmitic acid

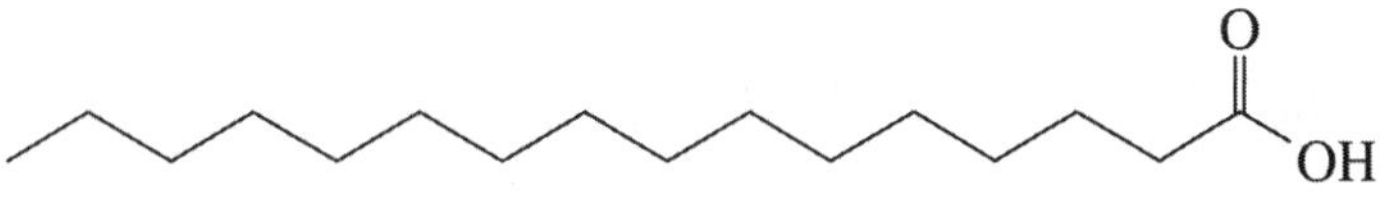

b. Oleic acid

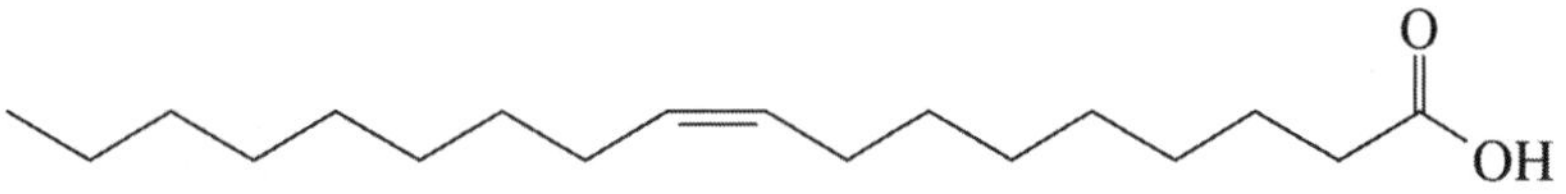

15.9 **a.** Lauric acid has only carbon–carbon single bonds; it is saturated.
 b. Linolenic acid has three carbon–carbon double bonds; it is unsaturated.
 c. Palmitoleic acid has one carbon–carbon double bond; it is unsaturated.
 d. Stearic acid has only carbon–carbon single bonds; it is saturated.

15.11 In a cis fatty acid, the hydrogen atoms are on the same side of the double bond, which produces a kink in the carbon chain. In a trans fatty acid, the hydrogen atoms are on opposite sides of the double bond, which gives a carbon chain without any kink.

15.13 In an omega-3 fatty acid, there is a double bond beginning at carbon 3, counting from the methyl group.
 In an omega-6 fatty acid, there is a double bond beginning at carbon 6, counting from the methyl group.

15.15 Arachidonic acid contains four double bonds and no side groups. In PGE_1, a part of the chain forms cyclopentane, and there are hydroxyl and ketone functional groups.

15.17 Prostaglandins raise or lower blood pressure, stimulate contraction and relaxation of smooth muscle, and may cause inflammation and pain.

15.19 Palmitic acid is the 16-carbon saturated fatty acid and myricyl alcohol has a 30-carbon chain.

$$CH_3-(CH_2)_{14}-\overset{\overset{\displaystyle O}{\|}}{C}-O-(CH_2)_{29}-CH_3$$

15.21 Triacylglycerols are composed of fatty acids and glycerol. In this case, the fatty acid is stearic acid, an 18-carbon saturated fatty acid.

$$
\begin{array}{l}
CH_2-O-\overset{\overset{\displaystyle O}{\|}}{C}-(CH_2)_{16}-CH_3 \\
\quad\quad\quad\quad\; \overset{\displaystyle O}{\|} \\
CH-O-\overset{}{C}-(CH_2)_{16}-CH_3 \\
\quad\quad\quad\quad\; \overset{\displaystyle O}{\|} \\
CH_2-O-\overset{}{C}-(CH_2)_{16}-CH_3
\end{array}
$$

15.23 Glyceryl tricaprate (tricaprin) has three capric acids (a 10-carbon saturated fatty acid) forming ester bonds with glycerol.

$$
\begin{array}{l}
CH_2-O-\overset{\overset{\displaystyle O}{\|}}{C}-(CH_2)_{8}-CH_3 \\
\quad\quad\quad\quad\; \overset{\displaystyle O}{\|} \\
CH-O-\overset{}{C}-(CH_2)_{8}-CH_3 \\
\quad\quad\quad\quad\; \overset{\displaystyle O}{\|} \\
CH_2-O-\overset{}{C}-(CH_2)_{8}-CH_3
\end{array}
$$

15.25 Safflower oil contains fatty acids with two or three double bonds; olive oil contains a large amount of oleic acid, which has only one (monounsaturated) double bond.

15.27 Although coconut oil comes from a plant source, it has large amounts of saturated fatty acids and small amounts of unsaturated fatty acids. Since coconut oil contains the same kinds of fatty acids as animal fat, coconut oil has a melting point similar to the melting point of animal fats.

15.29 Hydrogenation of an unsaturated triacylglycerol adds H_2 to each of the double bonds, producing a saturated triacylglycerol containing only carbon–carbon single bonds.

$$
\begin{array}{l}
CH_2-O-\overset{\overset{\displaystyle O}{\|}}{C}-(CH_2)_{7}-CH=CH-(CH_2)_{7}-CH_3 \\
\quad\quad\quad\quad\; \overset{\displaystyle O}{\|} \\
CH-O-\overset{}{C}-(CH_2)_{7}-CH=CH-(CH_2)_{7}-CH_3 \;\;+\;\; 3H_2 \\
\quad\quad\quad\quad\; \overset{\displaystyle O}{\|} \\
CH_2-O-\overset{}{C}-(CH_2)_{7}-CH=CH-(CH_2)_{7}-CH_3
\end{array}
\;\;\xrightarrow{\;\text{Ni}\;}\;\;
\begin{array}{l}
CH_2-O-\overset{\overset{\displaystyle O}{\|}}{C}-(CH_2)_{16}-CH_3 \\
\quad\quad\quad\quad\; \overset{\displaystyle O}{\|} \\
CH-O-\overset{}{C}-(CH_2)_{16}-CH_3 \\
\quad\quad\quad\quad\; \overset{\displaystyle O}{\|} \\
CH_2-O-\overset{}{C}-(CH_2)_{16}-CH_3
\end{array}
$$

15.31 Acid hydrolysis of a fat gives glycerol and the fatty acids. Basic hydrolysis (saponification) of fat gives glycerol and the salts of the fatty acids.

a.

$$
\begin{array}{l}
CH_2-O-\overset{\displaystyle O}{\overset{\|}{C}}-(CH_2)_{12}-CH_3 \\
CH-O-\overset{\displaystyle O}{\overset{\|}{C}}-(CH_2)_{12}-CH_3 \; + \; 3H_2O \; \xrightarrow{\;H^+\;} \\
CH_2-O-\overset{\displaystyle O}{\overset{\|}{C}}-(CH_2)_{12}-CH_3
\end{array}
$$

$$
\begin{array}{l}
CH_2-OH \\
CH-OH \; + \; 3HO-\overset{\displaystyle O}{\overset{\|}{C}}-(CH_2)_{12}-CH_3 \\
CH_2-OH
\end{array}
$$

b.

$$
\begin{array}{l}
CH_2-O-\overset{\displaystyle O}{\overset{\|}{C}}-(CH_2)_{12}-CH_3 \\
CH-O-\overset{\displaystyle O}{\overset{\|}{C}}-(CH_2)_{12}-CH_3 \; + \; 3NaOH \; \longrightarrow \\
CH_2-O-\overset{\displaystyle O}{\overset{\|}{C}}-(CH_2)_{12}-CH_3
\end{array}
$$

$$
\begin{array}{l}
CH_2-OH \\
CH-OH \; + \; 3Na^+\,{}^-O-\overset{\displaystyle O}{\overset{\|}{C}}-(CH_2)_{12}-CH_3 \\
CH_2-OH
\end{array}
$$

15.33 A triacylglycerol is composed of glycerol with three hydroxyl groups that form ester links with three long-chain fatty acids. In olestra, six to eight long-chain fatty acids form ester links with the hydroxyl groups on sucrose, a sugar. The olestra cannot be digested because our enzymes cannot break down the large olestra molecule.

15.35

$$
\begin{array}{l}
CH_2-O-\overset{\displaystyle O}{\overset{\|}{C}}-(CH_2)_{16}-CH_3 \\
CH-O-\overset{\displaystyle O}{\overset{\|}{C}}-(CH_2)_{16}-CH_3 \\
CH_2-O-\overset{\displaystyle O}{\overset{\|}{C}}-(CH_2)_{16}-CH_3
\end{array}
$$

15.37 A triacylglycerol consists of glycerol and three fatty acids. A glycerophospholipid consists of glycerol, two fatty acids, a phosphate group, and an amino alcohol.

15.39

$$
\begin{array}{l}
CH_2-O-\overset{\displaystyle O}{\overset{\|}{C}}-(CH_2)_{14}-CH_3 \\
CH-O-\overset{\displaystyle O}{\overset{\|}{C}}-(CH_2)_{14}-CH_3 \\
CH_2-O-\overset{\displaystyle O}{\underset{\underset{O^-}{|}}{\overset{\|}{P}}}-O-CH_2-CH_2-\overset{+}{N}H_3
\end{array}
$$

This is a cephalin.

15.41 This glycerophospholipid is a cephalin. It contains glycerol, oleic acid, stearic acid, phosphate, and ethanolamine.

15.43

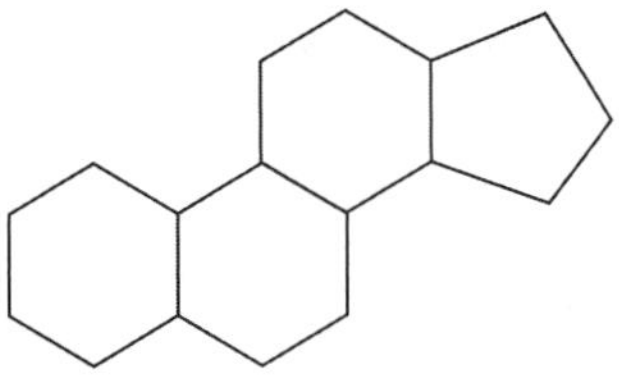

15.45 Bile salts act to emulsify fat globules, allowing the fat to be more easily digested by lipases.

15.47 Chylomicrons have a lower density than VLDLs. They pick up triacylglycerols from the intestine, whereas VLDLs transport triacylglycerols synthesized in the liver.

15.49 "Bad" cholesterol is the cholesterol carried by LDLs that can form deposits in the arteries called plaque, which narrows the arteries.

15.51 Both estradiol and testosterone contain the steroid nucleus and a hydroxyl group. Testosterone has a ketone group, a double bond, and two methyl groups. Estradiol has an aromatic ring, a hydroxyl group in place of the ketone, and a methyl group.

15.53 **d.** Testosterone is a male sex hormone.

15.55 The lipids in a cell membrane are glycerophospholipids with smaller amounts of cholesterol.

15.57 The function of the lipid bilayer in the cell membrane is to keep the cell contents separated from the outside environment and to allow the cell to regulate the movement of substances into and out of the cell.

15.59 The peripheral proteins in the membrane emerge on the inner or outer surface only, whereas the integral proteins extend through the membrane to both surfaces.

15.61 The carbohydrates attached to proteins and lipids on the surface of cells act as receptors for cell recognition and chemical messengers such as neurotransmitters.

15.63

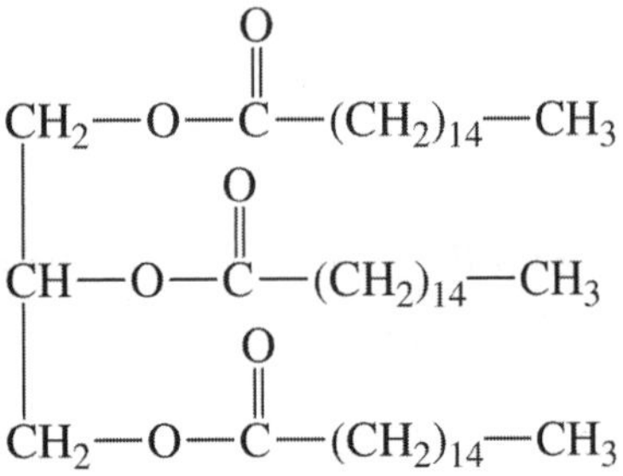

$$CH_2-O-\overset{\overset{\displaystyle O}{\|}}{C}-(CH_2)_{14}-CH_3$$
$$CH-O-\overset{\overset{\displaystyle O}{\|}}{C}-(CH_2)_{14}-CH_3 \qquad \text{Glyceryl tripalmitate (tripalmitin)}$$
$$CH_2-O-\overset{\overset{\displaystyle O}{\|}}{C}-(CH_2)_{14}-CH_3$$

15.65 **a.**

$$CH_2-O-\overset{\overset{\displaystyle O}{\|}}{C}-(CH_2)_7-CH=CH-CH_2-CH=CH-(CH_2)_4-CH_3$$
$$CH-O-\overset{\overset{\displaystyle O}{\|}}{C}-(CH_2)_7-CH=CH-(CH_2)_7-CH_3$$
$$CH_2-O-\overset{\overset{\displaystyle O}{\|}}{C}-(CH_2)_7-CH=CH-CH_2-CH=CH-(CH_2)_4-CH_3$$

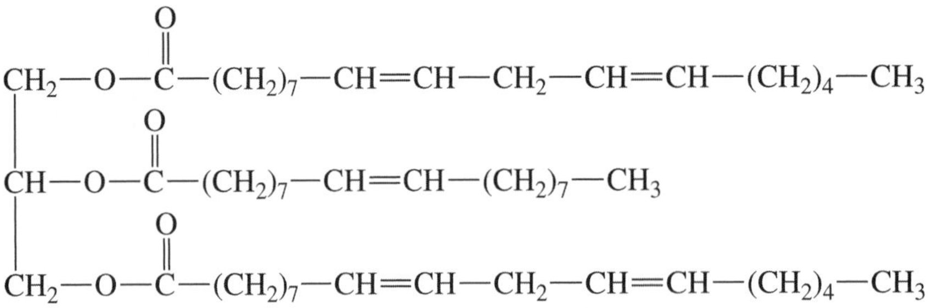

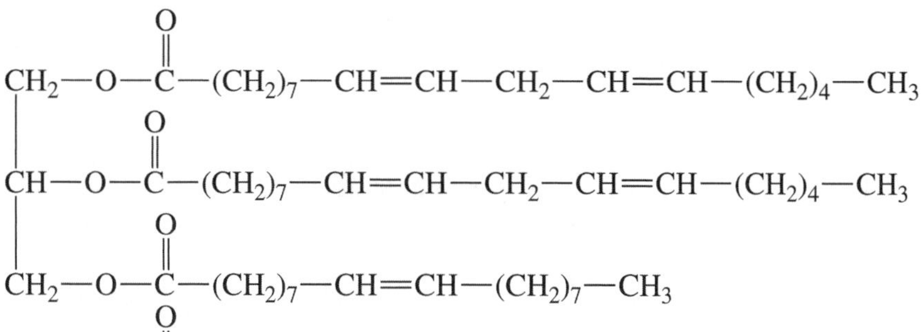

$$CH_2-O-\overset{\overset{O}{\|}}{C}-(CH_2)_7-CH=CH-CH_2-CH=CH-(CH_2)_4-CH_3$$

$$CH-O-\overset{\overset{O}{\|}}{C}-(CH_2)_7-CH=CH-CH_2-CH=CH-(CH_2)_4-CH_3$$

$$CH_2-O-\overset{\overset{O}{\|}}{C}-(CH_2)_7-CH=CH-(CH_2)_7-CH_3$$

b. 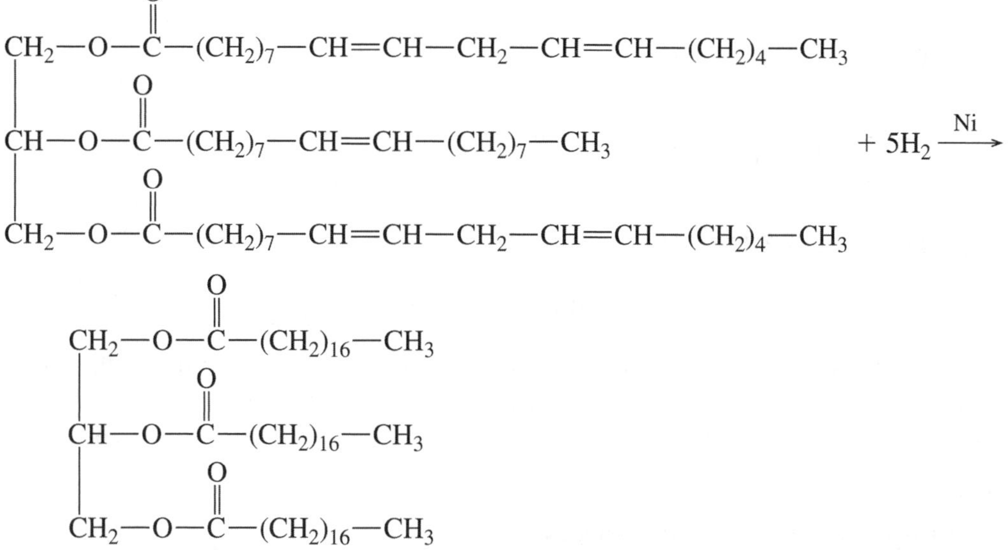
$$CH_2-O-\overset{\overset{O}{\|}}{C}-(CH_2)_7-CH=CH-CH_2-CH=CH-(CH_2)_4-CH_3$$

$$CH-O-\overset{\overset{O}{\|}}{C}-(CH_2)_7-CH=CH-(CH_2)_7-CH_3 \qquad +\ 5H_2 \xrightarrow{\text{Ni}}$$

$$CH_2-O-\overset{\overset{O}{\|}}{C}-(CH_2)_7-CH=CH-CH_2-CH=CH-(CH_2)_4-CH_3$$

$$CH_2-O-\overset{\overset{O}{\|}}{C}-(CH_2)_{16}-CH_3$$

$$CH-O-\overset{\overset{O}{\|}}{C}-(CH_2)_{16}-CH_3$$

$$CH_2-O-\overset{\overset{O}{\|}}{C}-(CH_2)_{16}-CH_3$$

15.67 Beeswax and carnauba wax are waxes. Vegetable oil and glyceryl tricaprate (tricaprin) are triacylglycerols.

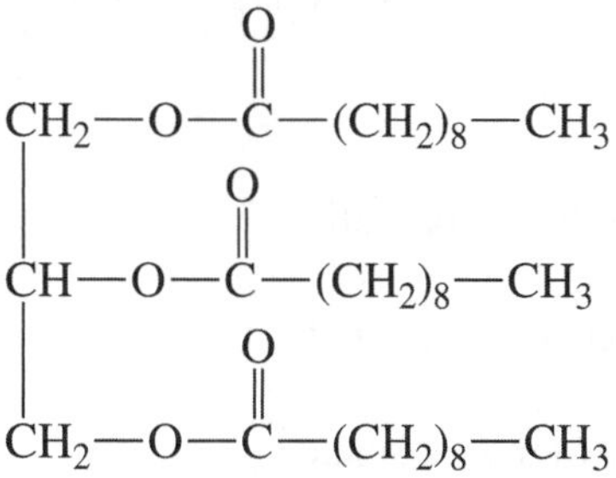

$$CH_2-O-\overset{\overset{O}{\|}}{C}-(CH_2)_8-CH_3$$

$$CH-O-\overset{\overset{O}{\|}}{C}-(CH_2)_8-CH_3 \qquad \text{Glyceryl tricaprate (tricaprin)}$$

$$CH_2-O-\overset{\overset{O}{\|}}{C}-(CH_2)_8-CH_3$$

15.69 **a.** A typical unsaturated fatty acid has a cis double bond.
b. A cis unsaturated fatty acid contains hydrogen atoms on the same side of each double bond. A trans unsaturated fatty acid has hydrogen atoms on the opposite sides of each double bond.

c. 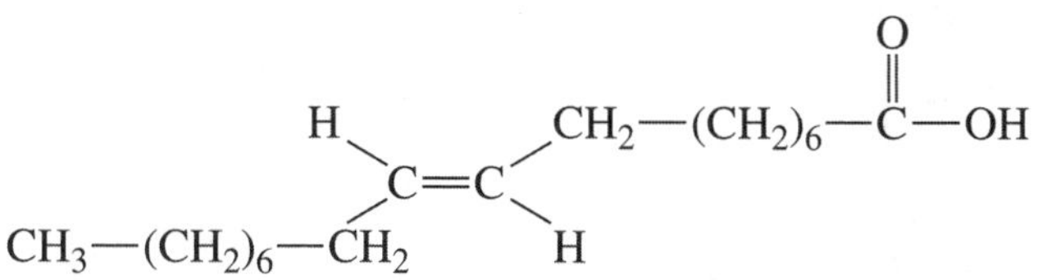 $CH_3-(CH_2)_6-CH_2$ ... $C=C$ ... $CH_2-(CH_2)_6-\overset{\overset{O}{\|}}{C}-OH$ (H, H)

15.71

$$CH_2-O-\overset{\overset{\displaystyle O}{\|}}{C}-(CH_2)_{16}-CH_3$$
$$|$$
$$CH-O-\overset{\overset{\displaystyle O}{\|}}{C}-(CH_2)_{16}-CH_3 \qquad \text{Glyceryl tristearate (tristearin)}$$
$$|$$
$$CH_2-O-\overset{\overset{\displaystyle O}{\|}}{C}-(CH_2)_{16}-CH_3$$

$$CH_2-O-\overset{\overset{\displaystyle O}{\|}}{C}-(CH_2)_{14}-CH_3$$
$$|$$
$$CH-O-\overset{\overset{\displaystyle O}{\|}}{C}-(CH_2)_{14}-CH_3 \qquad\qquad\qquad \text{Lecithin}$$
$$|$$
$$CH_2-O-\overset{\overset{\displaystyle O}{\|}}{\underset{\underset{\displaystyle O^-}{|}}{P}}-O-CH_2-CH_2-\overset{\overset{\displaystyle CH_3}{|}}{\underset{\underset{\displaystyle CH_3}{|}}{N^+}}-CH_3$$

15.73 Stearic acid (**k**) is a fatty acid. Sodium stearate (**e**) is a soap. Glyceryl tripalmitate (**d**), safflower oil (**f**), whale blubber (**g**), and adipose tissue (**h**) are triacylglycerols. Beeswax (**a**) is a wax. Lecithin (**c**) is a glycerophospholipid. Cholesterol (**b**), progesterone (**i**), and cortisone (**j**) are steroids.

15.75 **a.** Estrogen contains the steroid nucleus (**5**).
  b. Cephalin contains glycerol (**1**), fatty acids (**2**), phosphate (**3**), and an amino alcohol (**4**).
  c. Waxes contain fatty acid (**2**).
  d. Triacylglycerols contain glycerol (**1**) and fatty acids (**2**).
  e. Glycerophosholipids contain glycerol (**1**), fatty acids (**2**), phosphate (**3**), and an amino alcohol (**4**).

15.77 **a.** HDL (**4**) is known as "good" cholesterol.
  b. LDL (**3**) transports most of the cholesterol to the cells.
  c. Chylomicrons (**1**) carry triacylglycerols from the intestine to the fat cells.
  d. HDL (**4**) transports cholesterol to the liver.
  e. HDL (**4**) is the class of lipoprotein with the greatest abundance of protein.
  f. LDL (**3**) is known as "bad" cholesterol.
  g. VLDL (**2**) carries triacylglycerols synthesized in the liver to the muscles.
  h. Chylomicrons (**1**) are the lipoproteins with the lowest density.

15.79
$$CH_2-O-\overset{\overset{\displaystyle O}{\|}}{C}-(CH_2)_{16}-CH_3$$
$$|$$
$$CH-O-\overset{\overset{\displaystyle O}{\|}}{C}-(CH_2)_{16}-CH_3$$
$$|$$
$$CH_2-O-\overset{\overset{\displaystyle O}{\|}}{\underset{\underset{\displaystyle O^-}{|}}{P}}-O-CH_2-CH_2-\overset{+}{N}H_3$$

15.81 a. Adding NaOH will hydrolyze lipids such as glyceryl tristearate (tristearin), forming glycerol and salts of the fatty acids that are soluble in water and wash down the drain.

$$\mathbf{b.}\quad \underset{\displaystyle \begin{array}{c} \\[-2pt] \end{array}}{\overset{\displaystyle O}{\underset{\displaystyle \|}{}}}$$

b.
CH$_2$—O—C(=O)—(CH$_2$)$_{16}$—CH$_3$
|
CH—O—C(=O)—(CH$_2$)$_{16}$—CH$_3$ + 3NaOH →
|
CH$_2$—O—C(=O)—(CH$_2$)$_{16}$—CH$_3$

CH$_2$—OH
|
CH—OH   + 3Na$^+$ $^-$O—C(=O)—(CH$_2$)$_{16}$—CH$_3$
|
CH$_2$—OH

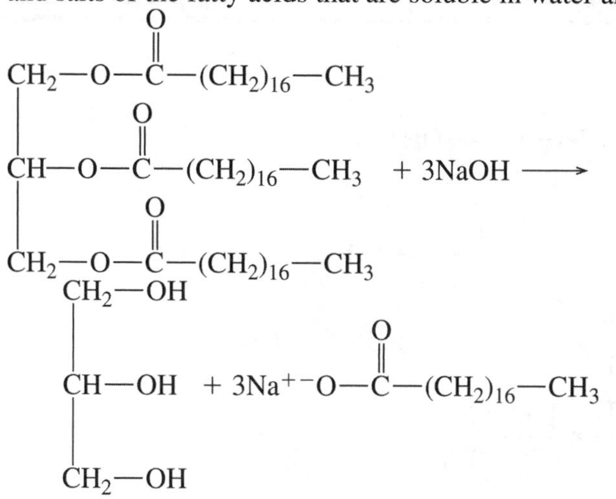

Glycerol   Sodium stearate

Answers to Combining Ideas from Chapters 13 to 15

CI.25 a.

$$HO-\overset{\overset{\displaystyle O}{\|}}{C}-\bigcirc-\overset{\overset{\displaystyle O}{\|}}{C}-O-CH_2-CH_2-OH$$

b.

$$HO-CH_2-CH_2-O-\overset{\overset{\displaystyle O}{\|}}{C}-\bigcirc-\overset{\overset{\displaystyle O}{\|}}{C}-O-CH_2-CH_2-OH$$

c. 1.7×10^9 lb PETE $\times \dfrac{1\text{ kg PETE}}{2.20\text{ lb PETE}} = 7.7\times10^8$ kg of PETE (2 SFs)

d. 1.7×10^9 lb PETE $\times \dfrac{454\text{ g PETE}}{1\text{ lb PETE}} \times \dfrac{1\text{ mL PETE}}{1.38\text{ g PETE}} \times \dfrac{1\text{ L PETE}}{1000\text{ mL PETE}}$

$= 5.6\times10^8$ L of PETE (2 SFs)

e. 5.6×10^8 L PETE $\times \dfrac{1\text{ landfill}}{2.7\times10^7\text{ L PETE}} = 21$ landfills (2 SFs)

CI.27 a.

$$\bigcirc-\overset{\overset{\displaystyle O}{\|}}{C}-\overset{\overset{\displaystyle CH_2-CH_3}{|}}{N}-CH_2-CH_3$$
$$CH_3$$

b. The molecular formula of DEET is $C_{12}H_{17}NO$.

c. Molar mass of DEET $(C_{12}H_{17}NO) = 12(12.0\text{ g}) + 17(1.01\text{ g}) + 1(14.0\text{ g}) + 1(16.0\text{ g})$

$= 191$ g/mole (3 SFs)

d. 6.0 fl oz $\times \dfrac{1\text{ qt}}{32\text{ fl oz}} \times \dfrac{946\text{ mL}}{1\text{ qt}} \times \dfrac{25\text{ g DEET}}{100\text{ mL solution}} = 44$ g of DEET (2 SFs)

e. 6.0 fl oz $\times \dfrac{1\text{ qt}}{32\text{ fl oz}} \times \dfrac{946\text{ mL}}{1\text{ qt}} \times \dfrac{25\text{ g DEET}}{100\text{ mL solution}} \times \dfrac{1\text{ mole DEET}}{191\text{ g DEET}}$

$\times \dfrac{6.02\times10^{23}\text{ molecules DEET}}{1\text{ mole DEET}} = 1.4\times10^{23}$ molecules of DEET (2 SFs)

CI.29 a. **A**, **B**, and **C** are all glucose units.

b. An α-1,6-glycosidic bond connects monosaccharides **A** and **B**.

c. An α-1,4-glycosidic bond connects monosaccharides **B** and **C**.

d. The structure is drawn is β-panose.

e. Panose is a reducing sugar because it has a free hydroxyl group on carbon 1 of structure **C**, which allows glucose (**C**) to form an aldehyde.

Amino Acids, Proteins, and Enzymes

16.1 **a.** Hemoglobin, which carries oxygen in the blood, is a transport protein.
    b. Collagen, which is a major component of tendons and cartilage, is a structural protein.
    c. Keratin, which is found in hair, is a structural protein.
    d. Amylases, which catalyze the breakdown of starch, are enzymes.

16.3 All α-amino acids contain a carboxylic acid group and an amino group on the alpha carbon.

16.5 **a.**

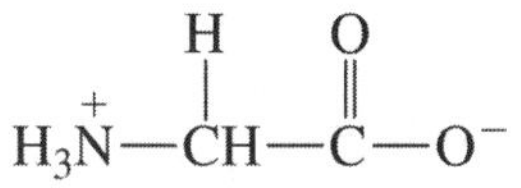

$$H_3\overset{+}{N}-\overset{\overset{\displaystyle H}{|}}{C}H-\overset{\overset{\displaystyle O}{\|}}{C}-O^-$$

b.

$$\begin{array}{c} CH_3 \\ | \\ HO-CH \\ | \\ H_3\overset{+}{N}-CH-\overset{\overset{\displaystyle O}{\|}}{C}-O^- \end{array}$$

c.

$$\begin{array}{c} O\diagdown\;\diagup O^- \\ C \\ | \\ CH_2 \\ | \\ CH_2 \\ | \\ H_3\overset{+}{N}-CH-\overset{\overset{\displaystyle O}{\|}}{C}-O^- \end{array}$$

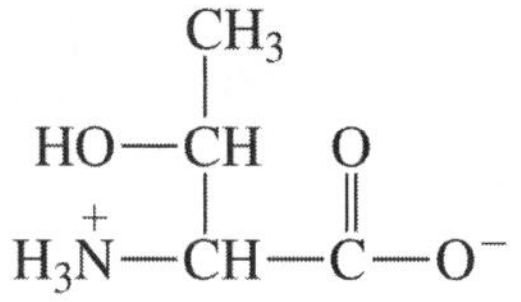

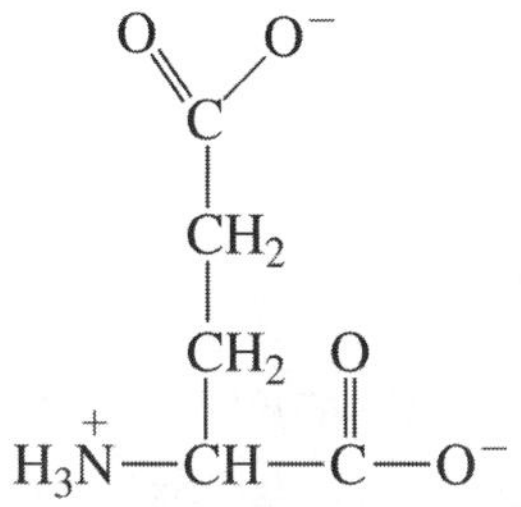

d.

$$\begin{array}{c} \text{(benzene ring)} \\ | \\ CH_2 \\ | \\ H_3\overset{+}{N}-CH-\overset{\overset{\displaystyle O}{\|}}{C}-O^- \end{array}$$

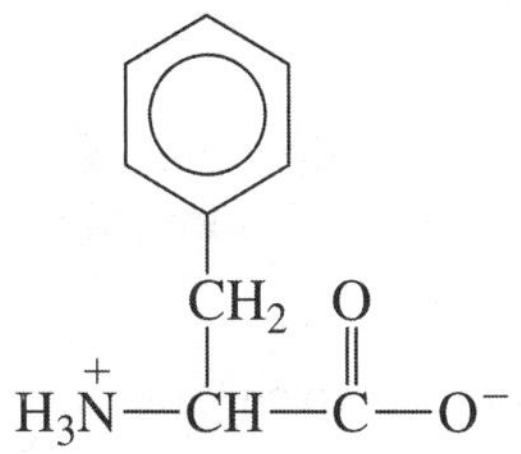

16.7 **a.** Glycine has an H as its R group which makes it a nonpolar amino acid.
    b. Threonine has an R group that contains the polar —OH group which makes threonine a polar neutral amino acid.
    c. Glutamic acid has an R group containing a polar carboxylic acid group. Glutamic acid is a polar acidic amino acid.
    d. Phenylalanine has an R group containing an aromatic ring; this makes phenylalanine a nonpolar amino acid.

16.9 The abbreviations of most amino acids are derived from the first three letters in the name.
    a. alanine
    b. valine
    c. lysine
    d. cysteine

16.11 At its isoelectric point (pI), an amino acid has an overall neutral (0) charge due to the ionized forms of both its amino ($-NH_3^+$) and carboxylic acid ($-COO^-$) groups. At a more acidic pH below its pI, the $-COO^-$ group gains H^+ to give $-COOH$, so a nonpolar amino acid would have an overall positive (1+) charge.

16.13 At low pH (highly acidic), the $-COO^-$ group of the zwitterion accepts H^+ and the amino acid has a positive charge overall.

a. 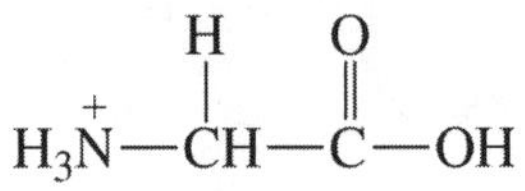
$$\overset{+}{H_3N}-\overset{\overset{\textstyle H}{|}}{CH}-\overset{\overset{\textstyle O}{\|}}{C}-OH$$

b.
$$\overset{+}{H_3N}-\overset{\overset{\textstyle SH}{|}\,\overset{\textstyle CH_2}{|}}{CH}-\overset{\overset{\textstyle O}{\|}}{C}-OH$$

c.
$$\overset{+}{H_3N}-\overset{\overset{\textstyle OH}{|}\,\overset{\textstyle CH_2}{|}}{CH}-\overset{\overset{\textstyle O}{\|}}{C}-OH$$

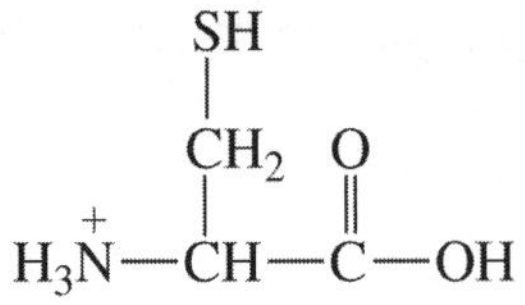

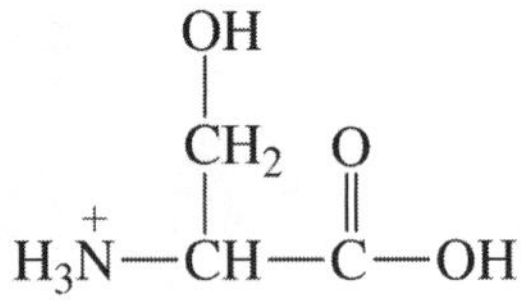

d. 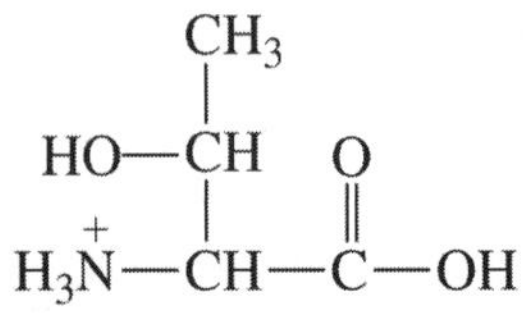
$$\overset{+}{H_3N}-\overset{\overset{\textstyle CH_3}{|}\,\overset{\textstyle HO-CH}{|}}{CH}-\overset{\overset{\textstyle O}{\|}}{C}-OH$$

16.15 a. When valine is in a solution with a pH above its pI, the $-NH_3^+$ group loses H^+ to form an uncharged amino group ($-NH_2$). The negative charge on the ionized carboxylate group ($-COO^-$) gives valine an overall negative charge (1−).

   b. When valine is in a solution with a pH below its pI, the $-COO^-$ group gains H^+ to form an uncharged carboxylic acid group ($-COOH$). The positive charge on the ionized ammonium group ($-NH_3^+$) gives valine an overall positive charge (1+).

   c. In a solution with a pH equal to its pI (6.0), valine is in its zwitterion form, containing both a carboxylate anion ($-COO^-$) and an ammonium cation ($-NH_3^+$), which give an overall charge of zero.

16.17 In a peptide, the amino acids are joined by peptide bonds (amide bonds). The first amino acid has a free amino group, and the last one has a free carboxyl group.

a. 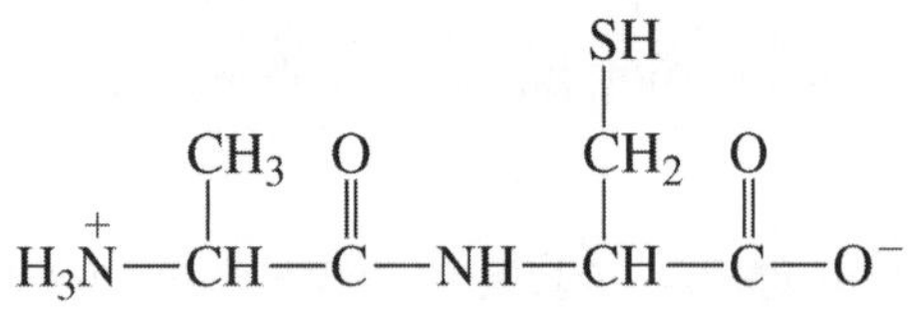
$$\overset{+}{H_3N}-\overset{\overset{\textstyle CH_3}{|}}{CH}-\overset{\overset{\textstyle O}{\|}}{C}-NH-\overset{\overset{\textstyle SH}{|}\,\overset{\textstyle CH_2}{|}}{CH}-\overset{\overset{\textstyle O}{\|}}{C}-O^-$$

Ala–Cys

b.

$$\text{H}_3\overset{+}{\text{N}}-\underset{\underset{\text{CH}_2}{\overset{\text{OH}}{|}}}{\text{CH}}-\underset{\text{O}}{\overset{\text{O}}{\text{C}}}-\text{NH}-\underset{\underset{\text{CH}_2}{|}}{\text{CH}}-\overset{\text{O}}{\text{C}}-\text{O}^-$$

Ser–Phe

c.

$$\text{H}_3\overset{+}{\text{N}}-\underset{\overset{\text{H}}{|}}{\text{CH}}-\overset{\text{O}}{\text{C}}-\text{NH}-\underset{\overset{\text{CH}_3}{|}}{\text{CH}}-\overset{\text{O}}{\text{C}}-\text{NH}-\underset{\overset{\text{CH}}{|}}{\text{CH}}-\overset{\text{O}}{\text{C}}-\text{O}^-$$

Gly–Ala–Val

d.

$$\text{H}_3\overset{+}{\text{N}}-\underset{\overset{\text{CH}}{|}}{\text{CH}}-\overset{\text{O}}{\text{C}}-\text{NH}-\underset{\overset{\text{CH}_2}{|}}{\text{CH}}-\overset{\text{O}}{\text{C}}-\text{NH}-\underset{\overset{\text{CH}_2}{|}}{\text{CH}}-\overset{\text{O}}{\text{C}}-\text{O}^-$$

Val–Ile–Trp

16.19 The possible primary structures of a tripeptide of one valine and two serines are Val–Ser–Ser, Ser–Val–Ser, and Ser–Ser–Val.

16.21 The primary structure remains unchanged and intact as hydrogen bonds form between carbonyl oxygen atoms and hydrogen atoms of amide groups in the polypeptide chain.

16.23 In the α helix, hydrogen bonds form between the carbonyl oxygen atom and the hydrogen atom of an amide group in the next turn of the helical chain. In the β-pleated sheet, hydrogen bonds occur between parallel peptides or across sections of a long polypeptide chain.

16.25 **a.** The two cysteines have — SH groups, which react to form a disulfide bond.
   b. Aspartic acid is acidic and lysine is basic; an ionic bond, or salt bridge, is formed between the — COO⁻ in the R group of Asp and the — NH₃⁺ in the R group of Lys.
   c. Serine has a polar — OH group that can form a hydrogen bond with the carboxyl group of aspartic acid.
   d. Two leucines have R groups that are hydrocarbons and nonpolar. They would have a hydrophobic interaction.

16.27 **a.** The R group of cysteine contains a — SH group that can form a disulfide cross-link.
   b. Leucine and valine will be found on the inside of the protein structure since they have nonpolar R groups that are hydrophobic.
   c. The cysteine and aspartic acid would be on the outside of the protein since they have R groups that are polar.
   d. The order of the amino acids (the primary structure) provides R groups that interact to determine the tertiary structure of the protein.

16.29 **a.** Disulfide bonds and ionic bonds join different sections of the protein chain to give a three-dimensional shape. Disulfide bonds and ionic bonds are important in the tertiary and quaternary structures.
   b. Peptide bonds join the amino acid building blocks in the primary structure of a polypeptide.

    c. Hydrogen bonds that hold adjacent polypeptide chains together are found in the secondary structures of β-pleated sheets of fibrous proteins and in the triple helices of collagen.

    d. In the secondary structure of α helices, hydrogen bonding occurs between amino acids in the same polypeptide to give a coiled shape to the protein.

16.31   **a.** Placing an egg in boiling water disrupts hydrogen bonds and hydrophobic interactions, which changes secondary and tertiary structures and causes loss of overall shape.

    b. Using an alcohol swab disrupts hydrophobic interactions and changes the tertiary structure of bacterial proteins on the surface of the skin.

    c. The heat from an autoclave will disrupt hydrogen bonds and hydrophobic interactions, which changes secondary and tertiary structures of the proteins in any bacteria present.

    d. Heat will cause changes in the secondary and tertiary structure of surrounding protein, which results in the formation of solid protein that helps to close the wound.

16.33   Chemical reactions in the body can occur without enzymes, but the rates are too slow at the relatively mild conditions of normal body temperature and pH. Catalyzed reactions, which are many times faster, provide the amounts of products needed by the cell at a particular time.

16.35   **a.** The reactant for the enzyme galactase is the sugar galactose.

    b. Lipase catalyzes the hydrolysis of lipids.

    c. The reactant for the enzyme aspartase is aspartic acid (aspartate).

16.37   **a.** A hydrolase enzyme would catalyze the hydrolysis of sucrose.

    b. An oxidoreductase enzyme would catalyze the addition of oxygen (oxidation).

    c. An isomerase enzyme would catalyze converting glucose to fructose.

    d. A transferase enzyme would catalyze moving an amino group from one molecule to another.

16.39   **a.** An enzyme (2) has a tertiary structure that recognizes the substrate.

    b. The combination of the enzyme and substrate is the enzyme–substrate complex (1).

    c. The substrate (3) has a structure that fits the active site of the enzyme.

16.41   **a.** The equation for an enzyme–catalyzed reaction is:

$$E + S \rightleftharpoons ES \longrightarrow E + P$$

E = enzyme, S = substrate, ES = enzyme–substrate complex, P = products

    b. The active site is a region or pocket within the tertiary structure of an enzyme that accepts the substrate, aligns the substrate for reaction, and catalyzes the reaction.

16.43   Isoenzymes are slightly different forms of an enzyme that catalyze the same reaction in different organs and tissues of the body.

16.45   A doctor might run tests for the enzymes CK, LDH, and AST to determine if the patient had a heart attack.

16.47   **a.** Decreasing the substrate concentration decreases the rate of reaction.

    b. Running the reaction at a pH below optimum pH will decrease the rate of reaction.

    c. Temperatures above 37 °C (optimum temperature) will denature the enzymes and decrease the rate of reaction.

    d. Increasing the enzyme concentration would increase the rate of reaction, as long as there are still free polypeptides to react ([S] < [E]).

16.49   From the graph, the optimum pH values for the enzymes are approximately: pepsin, pH 2; sucrase, pH 6; trypsin, pH 8.

16.51   **a.**   If the inhibitor has a structure similar to the substrate, the inhibitor is competitive.

        b.   If adding more substrate cannot reverse the effect of the inhibitor, the inhibitor is noncompetitive.

        c.   If the inhibitor competes with the substrate for the active site, it is a competitive inhibitor.

        d.   If the structure of the inhibitor is not similar to the substrate, the inhibitor is noncompetitive.

        e.   If adding more substrate reverses the inhibition, the inhibitor is competitive.

16.53   **a.**   Methanol has the condensed structural formula CH_3—OH, whereas ethanol is CH_3—CH_2—OH.

        b.   Ethanol has a structure similar to methanol and could compete for the active site.

        c.   Ethanol is a competitive inhibitor of methanol oxidation.

16.55   **a.**   Thiamine or vitamin B_1 is a cofactor (coenzyme), which is required by this enzyme for activity.

        b.   The Zn^{2+} is a cofactor, which is required by this enzyme for activity.

        c.   If the active form of an enzyme consists of just polypeptide chains, it is a simple enzyme.

16.57   **a.**   Pantothenic acid (vitamin B_5) is part of coenzyme A.

        b.   Tetrahydrofolate (THF) is a reduced form of folic acid.

        c.   Niacin (vitamin B_3) is a component of NAD^+.

16.59   **a.**   The oxidation of a glycol to an aldehyde and carboxylic acid is catalyzed by an oxidoreductase.

        b.   At high concentration, ethanol, which acts as a competitive inhibitor of ethylene glycol, would saturate the alcohol dehydrogenase enzyme to allow ethylene glycol to be removed from the body without producing oxalic acid.

16.61   **a.**   The amino acids in aspartame are aspartic acid and phenylalanine.

        b.   The dipeptide in aspartame would be named aspartylphenylalanine.

16.63   **a.**   Because a pH of 10.5 is more basic and above the pI of cysteine, the ($-NH_3^+$) group loses H^+ to form an uncharged amino group ($-NH_2$). The negative charge on the ionized carboxylate group ($-COO^-$) gives cysteine an overall negative charge ($1-$), as in diagram 1.

        b.   Since the pI of cysteine is 5.1, at that pH cysteine exists as a zwitterion with an overall charge of zero (diagram 3).

        c.   Because a pH of 1.8 is more acidic and below the pI of cysteine, the $-COO^-$ group gains H^+ to form an uncharged carboxylic acid group ($-COOH$). The positive charge on the ionized ammonium group ($-NH_3^+$) gives cysteine an overall positive charge ($1+$), as in diagram 2.

16.65   **a.**   Asparagine and serine are both polar neutral amino acids; their R groups can interact by hydrogen bonding.

        b.   Aspartic acid is a polar acidic amino acid, and lysine is a polar basic amino acid; their R groups interact by forming a salt bridge (ionic bond).

        c.   The polar neutral amino acid cysteine contains the $-SH$ group; two cysteines can form a disulfide bond.

        d.   Leucine and alanine are both nonpolar amino acids; their R groups have a hydrophobic interaction.

16.67 a.

$$\overset{+}{H_3N}\text{—CH—}\overset{O}{\overset{\|}{C}}\text{—NH—CH—}\overset{O}{\overset{\|}{C}}\text{—NH—CH—}\overset{O}{\overset{\|}{C}}\text{—O}^-$$

with R groups:
- CH₂—OH
- CH₂—CH₂—CH₂—CH₂—NH₃⁺
- CH₂—C(=O)(O⁻)

    b. This segment contains polar R groups, which would be found on the surface of a globular protein where they can hydrogen bond with water.

16.69 a. Yes, a combination of rice and garbanzo beans provides all the essential amino acids; garbanzo beans contain the lysine missing in rice.

    b. No, a combination of lima beans and cornmeal does not provide all the essential amino acids; both are deficient in the amino acid tryptophan.

    c. No, a combination of garbanzo beans and lima beans does not provide all the essential amino acids; both are deficient in the amino acid tryptophan.

16.71 a. The secondary structure of a protein depends on hydrogen bonds to form a helix or a pleated sheet; the tertiary structure is determined by the interactions of R groups such as disulfide bonds and salt bridges, and determines the three-dimensional shape of the protein.

    b. Nonessential amino acids can be synthesized by the body, but essential amino acids must be supplied by the diet.

    c. Polar amino acids have hydrophilic R groups, while nonpolar amino acids have hydrophobic R groups.

    d. Dipeptides contain two amino acids, whereas tripeptides contain three amino acids.

16.73 Glutamic acid is a polar acidic amino acid, whereas proline is nonpolar. The polar acidic R group of glutamic acid would be located on the outside surface of the protein where it would form hydrophilic interactions with water. However, the nonpolar proline would move to the hydrophobic center of the protein. In addition, if the glutamic acid were involved in a salt bridge, its replacement by nonpolar proline would remove an important cross-link in stabilizing tertiary structure.

16.75 In chemical laboratories, reactions are often run at high temperatures using catalysts that are strong acids or bases. Enzymes, which function at physiological temperatures and pH, are denatured rapidly if high temperatures or acids or bases are used.

16.77 a. The reactant is lactose, and the products are glucose and galactose.

    b.

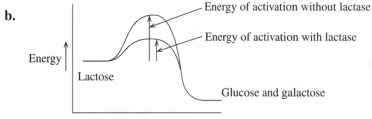

    c. By lowering the energy of activation, the enzyme furnishes a lower energy pathway by which the reaction can take place.

16.79 **a.** The disaccharide lactose is a substrate (S).
      b. The suffix *ase* in lipase indicates that it is an enzyme (E).
      c. The suffix *ase* in urease indicates that it is an enzyme (E).
      d. Trypsin is an enzyme which hydrolyzes polypeptides (E).
      e. Pyruvate is a substrate (S).
      f. The suffix *ase* in transaminase indicates that it is an enzyme (E).

16.81 **a.** Urea is the substrate of urease.
      b. Succinate is the substrate of succinate dehydrogenase.
      c. Aspartate is the substrate of aspartate transaminase.
      d. Tyrosine is the substrate of tyrosinase.

16.83 Sucrose fits the shape of the active site in sucrase, but lactose does not.

16.85 A heart attack may be the cause. Normally the enzymes LDH and CK are present only in low levels in the blood.

16.87 **a.** An enzyme is saturated if adding more substrate does not increase the rate of reaction.
      b. When increasing the substrate concentration increases the rate of reaction, the enzyme is not saturated.

16.89 **a.** The Mg^{2+} is a cofactor, which is required by this enzyme for activity.
      b. A protein that is catalytically active as a tertiary protein structure is a simple enzyme.
      c. Folic acid is a cofactor (coenzyme), which is required by this enzyme for activity.

16.91 **a.** Since the pI of serine is 5.7, at that pH serine exists as a zwitterion with an overall charge of zero (0).
      b. Because a pH of 2.0 is more acidic and below the pI of threonine (5.6), the $-COO^-$ group gains H^+ to form an uncharged carboxylic acid group ($-COOH$). The positive charge on the ionized ammonium group ($-NH_3^+$) gives threonine an overall positive charge (1+).
      c. Because a pH of 3.0 is more acidic and below the pI of isoleucine (6.0), the $-COO^-$ group gains H^+ to form an uncharged carboxylic acid group ($-COOH$). The positive charge on the ionized ammonium group ($-NH_3^+$) gives isoleucine an overall positive charge (1+).
      d. Because a pH of 9.0 is more basic and above the pI of leucine (6.0), the $-NH_3^+$ group loses H^+ to form an uncharged amino group ($-NH_2$). The negative charge on the ionized carboxylate group ($-COO^-$) gives leucine an overall negative charge (1−).

16.93 **a.** Valine has a nonpolar R group which would be found in hydrophobic regions.
      b. Lysine and aspartic acid have polar R groups which would be found in hydrophilic regions.
      c. Lysine and aspartic acid have polar R groups containing $-COO^-$ and $-NH_3^+$ groups which can form hydrogen bonds.
      d. Lysine is a polar basic amino acid, and aspartic acid is a polar acidic amino acid; they can form salt bridges.

17
Nucleic Acids and Protein Synthesis

17.1 Purine bases (e.g., adenine, guanine) have a double-ring structure; pyrimidines (e.g., cytosine, thymine, uracil) have a single ring.
   a. Thymine is a pyrimidine base.
   b. This single-ring base is the pyrimidine cytosine.

17.3 DNA contains two purines, adenine (A) and guanine (G), and two pyrimidines, cytosine (C) and thymine (T). RNA contains the same bases, except thymine (T) is replaced by the pyrimidine uracil (U).
   a. Thymine is a base present in DNA.
   b. Cytosine is a base present in both DNA and RNA.

17.5 Nucleotides contain a base, a sugar, and a phosphate group. The nucleotides found in DNA would all contain the sugar deoxyribose. The four nucleotides are deoxyadenosine-5′-monophosphate (dAMP), deoxyguanosine-5′-monophosphate (dGMP), deoxycytidine-5′-monophosphate (dCMP), and deoxythymidine-5′-monophosphate (dTMP).

17.7   **a.** Adenosine is a nucleoside found in RNA.
   b. Deoxycytidine is a nucleoside found in DNA.
   c. Uridine is a nucleoside found in RNA.
   d. Cytidine-5′-monophosphate is a nucleotide found in RNA.

17.9

17.11 The nucleotides in nucleic acid polymers are held together by phosphodiester bonds between the 3′—OH of a sugar (ribose or deoxyribose) and the phosphate group on the 5′-carbon of another sugar.

17.13

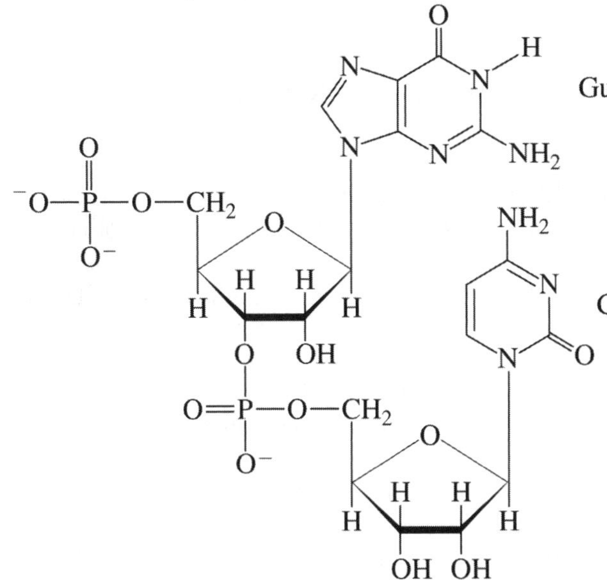

Guanosine (G)

Cytidine (C)

17.15 The two DNA strands are held together by hydrogen bonds between the complementary bases in each strand.

17.17 **a.** Since T pairs with A, if one strand of DNA has the sequence
— A — A — A — A — A — A —, the second strand would be
— T — T — T — T — T — T —.

b. Since C pairs with G, if one strand of DNA has the sequence
— G — G — G — G — G — G —, the second strand would be
— C — C — C — C — C — C —.

c. Since T pairs with A, and C pairs with G, if one strand of DNA has the sequence
— A — G — T — C — C — A — G — G — T —, the second strand would be
— T — C — A — G — G — T — C — C — A —.

d. Since T pairs with A, and C pairs with G, if one strand of DNA has the sequence
— C — T — G — T — A — T — A — C — G — T — T — A —, the second strand would be
— G — A — C — A — T — A — T — G — C — A — A — T —.

17.19 The three types of RNA are messenger RNA (mRNA), ribosomal RNA (rRNA), and transfer RNA (tRNA).

17.21 In transcription, the sequence of nucleotides on a DNA template strand is used to produce the base sequences of a messenger RNA. The DNA unwinds, and one strand is copied as complementary bases are placed in the mRNA molecule. In RNA, U (uracil) is paired with A in DNA.

17.23 To form mRNA, the bases in the DNA template strand are paired with their complementary bases: G with C, C with G, T with A, and A with U. The strand of mRNA would have the following sequence:
— G — G — C — U — U — C — C — A — A — G — U — G —.

17.25 A codon is a three-base sequence (triplet) in mRNA that codes for a specific amino acid in a protein.

17.27 **a.** The codon CUU in mRNA codes for the amino acid leucine (Leu).
b. The codon UCA in mRNA codes for the amino acid serine (Ser).
c. The codon GGU in mRNA codes for the amino acid glycine (Gly).
d. The codon AGG in mRNA codes for the amino acid arginine (Arg).

17.29 At the beginning of an mRNA, the codon AUG signals the start of protein synthesis; thereafter, the AUG codon specifies the amino acid methionine.

17.31 A codon is a base triplet in the mRNA template. An anticodon is the complementary triplet on a tRNA for a specific amino acid.

17.33 **a.** The codons ACC, ACA, and ACU in mRNA all code for threonine: — Thr — Thr — Thr — .
   b. The codon UUC codes for phenylalanine, and CCG and CCA both code for proline:
      — Phe — Pro — Phe — Pro — .
   c. The codon UAC codes for tyrosine, GGG for glycine, AGA for arginine, and UGU for cysteine: — Tyr — Gly — Arg — Cys — .

17.35 The new amino acid is joined by a peptide bond to the growing peptide chain. The ribosome moves to the next codon, which attaches to a tRNA carrying the next amino acid.

17.37 **a.** The mRNA sequence would be: — CGA — AAA — GUU — UUU — .
   b. The tRNA triplet anticodons would be: GCU, UUU, CAA, and AAA.
   c. From the table of mRNA codons, the amino acids would be: — Arg — Lys — Val — Phe — .

17.39 In a substitution mutation, a base in DNA is replaced by a different base.

17.41 In a frameshift mutation, a base is lost or gained, which changes the remaining sequence of nucleotides in the codons and therefore the amino acids in the remaining polypeptide chain.

17.43 The normal triplet TTT in DNA forms a codon AAA in mRNA. AAA codes for lysine. The mutation TTC in DNA forms a codon AAG in mRNA, which also codes for lysine. Thus, there is no effect on the amino acid sequence.

17.45 **a.** — Thr — Ser — Arg — Val — is the amino acid sequence produced by normal DNA.
   b. — Thr — Thr — Arg — Val — is the amino acid sequence produced by a mutation.
   c. — Thr — Ser — Gly — Val — is the amino acid sequence produced by a mutation.
   d. — Thr — STOP; protein synthesis would terminate early. If this mutation occurs early in the formation of the polypeptide, the resulting protein will probably be nonfunctional.
   e. The new protein will contain the sequence — Asp — Ile — Thr — Gly — .
   f. The new protein will contain the sequence — His — His — Gly — .

17.47 **a.** Both codons GCC and GCA code for alanine.
   b. A vital ionic cross-link in the tertiary structure of hemoglobin cannot be formed when the polar glutamic acid is replaced by valine, which is nonpolar. The resulting hemoglobin is malformed and less capable of carrying oxygen.

17.49 A virus contains either DNA or RNA, but not both, inside a protein coat.

17.51 **a.** An RNA-containing virus must make viral DNA from the RNA to produce a protein coat, allowing the virus to replicate and leave the cell to infect new cells.
   b. A virus that uses reverse transcription is a retrovirus.

17.53 Nucleoside analogs such as AZT and ddI are similar to the nucleosides required to make viral DNA in reverse transcription. When they are incorporated into viral DNA, the lack of a hydroxyl group on the $3'$-carbon in the sugar prevents the formation of the sugar-phosphate bonds and stops the replication of the virus.

17.55 a.

| A | G | G | T | C | G | C | C | T | Parent strand |

| T | C | C | A | G | C | G | G | A | New strand |

b.

| A | G | G | U | C | G | C | C | U |

c. (Arg)—(Ser)—(Pro)

17.57 DNA contains two purines, adenine (A) and guanine (G), and two pyrimidines, cytosine (C) and thymine (T). RNA contains the same bases, except thymine (T) is replaced by the pyrimidine uracil (U).
  a. Cytosine is a pyrimidine base.
  b. Adenine is a purine base.
  c. Uracil is a pyrimidine base.
  d. Thymine is a pyrimidine base.
  e. Guanine is a purine base.

17.59 a. Deoxythymidine is a nucleoside containing the base thymine and the sugar deoxyribose.
  b. Adenosine contains the base adenine and the sugar ribose.
  c. Cytidine contains the base cytosine and the sugar ribose.
  d. Deoxyguanosine contains the base guanine and the sugar deoxyribose.

17.61 Thymine and uracil are both pyrimidines, but thymine has a methyl group on carbon 5.

17.63 Both RNA and DNA are polymers of nucleotides connected through phosphodiester bonds between alternating sugar and phosphate groups with bases extending out from each sugar.

17.65 a. —C—T—G—A—A—T—C—C—G—
  b. —A—C—G—T—T—T—G—A—T—C—G—A—
  c. —T—A—G—C—T—A—G—C—T—A—G—C—

17.67 a. Transfer RNA (tRNA) is the smallest type of RNA.
  b. Ribosomal RNA (rRNA) makes up the highest percentage of RNA in the cell.
  c. Messenger RNA (mRNA) carries genetic information from the nucleus to the ribosomes.

17.69 a. ACU, ACC, ACA, and ACG are all codons for the amino acid threonine.
  b. UCU, UCC, UCA, UCG, AGU, and AGC are all codons for the amino acid serine.
  c. UGU and UGC are codons for the amino acid cysteine.

17.71 a. AAG codes for lysine.
  b. AUU codes for isoleucine.
  c. CGA codes for arginine.

17.73 Using the genetic code, the codons indicate the following amino acid sequence:
  START—Tyr—Gly—Gly—Phe—Leu—STOP

17.75 The anticodon on tRNA consists of the three complementary bases to the codon in mRNA.
  a. UCG
  b. AUA
  c. GGU

17.77 There are 9 amino acids in the nonapeptide oxytocin. The codon for each amino acid contains three nucleotides, plus a start and stop triplet, which makes a minimum total of 33 nucleotides.

17.79 **a.** Because A bonds with T, T is also 28%. Thus the sum of A + T is 56%, which leaves 44% divided equally between G and C, 22% G and 22% C.

       b. Because C bonds with G, G is also 20%. Thus the sum of C + G is 40%, which leaves 60% divided equally between A and T, 30% A and 30% T.

17.81 A DNA virus attaches to a cell and injects viral DNA that uses the host cell to produce copies of DNA to make viral RNA. A retrovirus injects viral RNA from which complementary DNA is produced by reverse transcription.

18
Metabolic Pathways and Energy Production

18.1 The digestion of polysaccharides takes place in stage 1.

18.3 In metabolism, a catabolic reaction breaks apart large molecules, releasing energy.

18.5 The hydrolysis of the phosphodiester bond (P — O — P) in ATP releases energy that is sufficient for energy-requiring processes in the cell.

18.7 Hydrolysis is the main reaction involved in the digestion of carbohydrates.

18.9 The bile salts emulsify fat to give small fat globules for lipase hydrolysis.

18.11 The digestion of proteins begins in the stomach and is completed in the small intestine.

18.13 In biochemical systems, oxidation is usually accompanied by gain of oxygen or loss of hydrogen. Loss of oxygen or gain of hydrogen usually accompanies reduction.
    a. The reduced form of NAD^+ is abbreviated NADH.
    b. The oxidized form of $FADH_2$ is abbreviated FAD.
    c. The coenzyme that participates in the formation of a carbon–carbon double bond is FAD.

18.15 **a.** Pantothenic acid is a component of coenzyme A.
    b. Niacin is the vitamin component of NAD^+.
    c. Ribitol is the sugar alcohol that is a component of riboflavin in FAD.

18.17 Glucose is the starting compound of glycolysis.

18.19 In the initial steps of glycolysis, ATP molecules are required to add phosphate groups to glucose (phosphorylation reactions).

18.21 ATP is produced directly in glycolysis in two places. In reaction 7, phosphate from 1,3-bisphosphoglycerate is transferred to ADP and yields ATP. In reaction 10, phosphate from phosphoenolpyruvate is transferred directly to ADP.

18.23 **a.** In the phosphorylation of glucose to glucose-6-phosphate, one ATP is required.
    b. One NADH is produced in the conversion of each glyceraldehyde-3-phosphate to 1,3-bisphosphoglycerate.
    c. When glucose is converted to pyruvate, two ATPs and two NADHs are produced.

18.25 A cell converts pyruvate to acetyl CoA only under aerobic conditions; there must be sufficient oxygen available.

18.27 The oxidation of pyruvate converts NAD^+ to NADH and produces acetyl CoA and CO_2.

$$CH_3 - \overset{\displaystyle O}{\overset{\|}{C}} - COO^- + NAD^+ + HS - CoA \rightarrow CH_3 - \overset{\displaystyle O}{\overset{\|}{C}} - S - CoA + CO_2 + NADH + H^+$$

Pyruvate   Acetyl CoA

18.29 When pyruvate is reduced to lactate, the NAD^+ is used to oxidize glyceraldehyde-3-phosphate, regenerating NADH, which allows glycolysis to continue.

18.31 One turn of the citric acid cycle converts 1 acetyl CoA to $2CO_2$, $3NADH + 3H^+$, $FADH_2$, GTP (ATP), and HS—CoA.

18.33 The reactions in steps 3 and 4 involve oxidation and decarboxylation, which reduces the length of the carbon chain by one carbon in each reaction.

18.35 NAD^+ is reduced by the oxidation reactions 3, 4, and 8 of the citric acid cycle.

18.37 In reaction 5, GDP undergoes a direct phosphate transfer to yield GTP, which converts ADP to ATP and regenerates GDP for the citric acid cycle.

18.39 **a.** The six-carbon compounds in the citric acid cycle are citrate and isocitrate.
  b. Decarboxylation reactions remove carbon atoms as CO_2, which reduces the number of carbon atoms in the chain (reactions 3 and 4).
  c. The one five-carbon compound is α-ketoglutarate.
  d. Several reactions are oxidation reactions: isocitrate $\rightarrow$ α-ketoglutarate; α-ketoglutarate $\rightarrow$ succinyl CoA; succinate $\rightarrow$ fumarate; malate $\rightarrow$ oxaloacetate.
  e. Secondary alcohols are oxidized in reactions 3 and 8.

18.41 NADH and $FADH_2$ produced in glycolysis, oxidation of pyruvate, and the citric acid cycle provide the electrons for electron transport.

18.43 The mobile carrier coenzyme Q (or Q) transfers electrons from complex I to complex III.

18.45 When NADH transfers electrons to complex I, NAD^+ is produced.

18.47 In oxidative phosphorylation, the energy from the oxidation reactions in electron transport is used to drive ATP synthesis.

18.49 Protons return to a lower energy in the mitochondrial matrix by passing through ATP synthase, which releases energy to drive the synthesis of ATP.

18.51 The reduced coenzymes NADH and $FADH_2$ from glycolysis and the citric acid cycle transfer electrons to the electron transport system, which generates energy to drive the synthesis of ATP.

18.53 **a.** 3 ATP are produced by the oxidation of NADH in electron transport.
  b. 6 ATP are produced in glycolysis when glucose degrades to 2 pyruvate molecules.
  c. 6 ATP are produced when 2 pyruvate are oxidized to 2 acetyl CoA and $2CO_2$.
  d. 12 ATP are produced when acetyl CoA is broken down to $2CO_2$ in the citric acid cycle.

18.55 Fatty acids are activated in the cytosol at the outer mitochondrial membrane.

18.57 **a.** The β oxidation of a chain of 8 carbon atoms produces 4 acetyl CoA units.
  b. A C_8 fatty acid will go through 3 β-oxidation cycles.
  c. 48 ATP from 4 acetyl CoA (citric acid cycle) + 9 ATP from 3 NADH + 6 ATP from 3 $FADH_2$ − 2 ATP (activation) = 63 − 2 = 61 ATP

18.59 The energy from the oxidation of unsaturated fatty acids is slightly less than the energy from saturated fatty acids. For simplicity, we will assume that the total ATP production is the same for both saturated and unsaturated fatty acids.

    a. The β oxidation of a chain of 18 carbon atoms produces 9 acetyl CoA units.

    b. A C_{18} fatty acid will go through 8 β-oxidation cycles.

    c. 108 ATP from 9 acetyl CoA (citric acid cycle) + 24 ATP from 8 NADH + 16 ATP from 8 $FADH_2$ − 2 ATP (activation) = 148 − 2 = 146 ATP

18.61 Ketone bodies form in the body when excess acetyl CoA results from the breakdown of large amounts of fat. This occurs in starvation, fasting, low-carbohydrate diets, and diabetes.

18.63 High levels of ketone bodies lead to ketosis, a condition characterized by acidosis (a drop in blood pH values), excessive urination, and strong thirst.

18.65 In transamination, an amino group is transferred from an amino acid to an α-keto acid, creating a new amino acid and a new α-keto acid.

    a.
$$H-\overset{\overset{\displaystyle O}{\|}}{C}-COO^-$$

    b.
$$CH_3-\overset{\overset{\displaystyle O}{\|}}{C}-COO^-$$

18.67 NH_4^+ is toxic if allowed to accumulate in the body.

18.69   **a.** The three-carbon atom structure of alanine is converted to pyruvate.

    b. The four-carbon structure of aspartate is converted to oxaloacetate.

    c. Tyrosine can be converted to the four-carbon compounds acetoacetate and fumarate.

    d. The five-carbon structure from glutamine can be converted to α-ketoglutarate.

18.71 Lauric acid, $CH_3-(CH_2)_{10}-COOH$, is a C_{12} fatty acid ($C_{12}H_{24}O_2$).

    a. and b.
$$CH_3-(CH_2)_8-\underset{\beta}{CH_2}-\underset{\alpha}{CH_2}-\overset{\overset{\displaystyle O}{\|}}{C}-S-CoA$$

    c. 6 acetyl CoA units are produced.

    d. 5 cycles of β oxidation are needed.

    e.

activation	$\rightarrow$	−2 ATP
6 acetyl CoA $\times$ 12 ATP/acetyl CoA	$\rightarrow$	72 ATP
5 $FADH_2$ $\times$ 2 ATP/$FADH_2$	$\rightarrow$	10 ATP
5 NADH $\times$ 3 ATP/NADH	$\rightarrow$	15 ATP
	Total	95 ATP

18.73   **a.** Glucose is a product of the digestion of carbohydrate.

    b. Fatty acid is a product of the digestion of fat.

    c. Maltose is a product of the digestion of carbohydrate.

    d. Glycerol is a product of the digestion of fat.

    e. Amino acids are the products of the digestion of protein.

    f. Dextrins are produced by the digestion of carbohydrate.

18.75 $ATP + H_2O \rightarrow ADP + P_i + 7.3$ kcal/mole

18.77 Lactose undergoes digestion in the mucosal cells of the small intestine to yield galactose and glucose.

18.79 Glucose is the reactant, and pyruvate is the product of glycolysis.

18.81 Pyruvate is converted to lactate when oxygen is not present in the cell (anaerobic conditions) to regenerate NAD^+ for glycolysis.

18.83 The oxidation reactions of the citric acid cycle produce a source of reduced coenzymes for electron transport and ATP synthesis.

18.85 The oxidized coenzymes NAD^+ and FAD needed for the citric acid cycle are regenerated by the electron transport system.

18.87 In the chemiosmotic model, energy released as protons flow through ATP synthase and then back to the mitochondrial matrix is utilized for the synthesis of ATP.

18.89 The oxidation of glucose to pyruvate by glycolysis produces 6 ATPs. 2 ATPs are formed by direct phosphorylation along with 2 NADHs. Because the 2 NADHs are produced in the cytosol, the electrons are transferred to form 2 $FADH_2$s, which produces an additional 4 ATPs. The oxidation of glucose to CO_2 and H_2O produces 36 ATPs.

18.91 **a.** $24 \text{ h} \times \dfrac{60 \text{ min}}{1 \text{ h}} \times \dfrac{60 \text{ s}}{1 \text{ min}} \times \dfrac{2 \times 10^6 \text{ ATP}}{1 \text{ s cell}} \times 10^{13} \text{ cells} \times \dfrac{1 \text{ mole ATP}}{6.02 \times 10^{23} \text{ ATP}} \times \dfrac{7.3 \text{ kcal}}{1 \text{ mole ATP}}$

$= 21 \text{ kcal (2 SFs)}$

b. $24 \text{ h} \times \dfrac{60 \text{ min}}{1 \text{ h}} \times \dfrac{60 \text{ s}}{1 \text{ min}} \times \dfrac{2 \times 10^6 \text{ ATP}}{1 \text{ s cell}} \times 10^{13} \text{ cells} \times \dfrac{1 \text{ mole ATP}}{6.02 \times 10^{23} \text{ ATP}} \times \dfrac{507 \text{ g ATP}}{1 \text{ mole ATP}}$

$= 1500 \text{ g of ATP (2 SFs)}$

18.93 **a.** Glucose forming two pyruvates in glycolysis yields 6 ATP/glucose.
b. Pyruvate forming acetyl CoA yields 3 ATP/pyruvate.
c. Glucose forming two acetyl CoAs would yield 12 ATP/glucose.
d. Acetyl CoA going through one turn of the citric acid cycle yields 12 ATP/acetyl CoA.
e. The complete oxidation of caproic acid (C_6) yields 44 ATP/C_6 acid.
f. The oxidation of NADH to NAD^+ yields 3 ATP/NADH.
g. The oxidation of $FADH_2$ to FAD yields 2 ATP/$FADH_2$.

18.95 **a.** The disaccharide maltose will produce more ATP per mole than the monosaccharide glucose.
b. The 18C fatty acid stearic acid will produce more ATP per mole than the 14C fatty acid myristic acid.
c. The six-carbon molecule glucose will produce more ATP per mole than two two-carbon molecules of acetyl CoA.
d. The 8C fatty acid caprylic acid will produce more ATP per mole than the six-carbon molecule glucose.
e. The six-carbon compound citrate occurs earlier in the citric acid cycle than the four-carbon compound succinate and will produce more ATP per mole in one turn of the cycle.

Answers to Combining Ideas from Chapters 16 to 18

CI.31 **a.** An alpha-galactosidase is a hydrolase.

b. The substrate of the enzyme is the α-D-galactose unit in polysaccharides.

c. You should not heat or cook beano because high temperatures will denature the hydrolase enzyme so it no longer functions.

CI.33 **a.** Succinate is part of the citric acid cycle.

b. QH_2 is part of electron transport.

c. FAD is part of both the citric acid cycle and electron transport.

d. *cyt c* is part of electron transport.

e. H_2O is part of both the citric acid cycle and electron transport.

f. Malate is part of the citric acid cycle.

g. NAD^+ is part of both the citric acid cycle and electron transport.

CI.35 **a.** The components of acetyl CoA are aminoethanethiol, pantothenic acid, and phosphorylated adenosine diphosphate.

b. Coenzyme A carries an acetyl group to the citric acid cycle for oxidation.

c. The acetyl group links to the S atom in the aminoethanethiol part of CoA.

d. Molar mass of acetyl CoA ($C_{23}H_{38}N_7O_{17}P_3S$)

$= 23(12.0 \text{ g}) + 38(1.01 \text{ g}) + 7(14.0 \text{ g}) + 17(16.0 \text{ g}) + 3(31.0 \text{ g}) + 1(32.1 \text{ g}) = 809 \text{ g/mole}$

CI.37 **a.**

b. Molar mass of glyceryl tripalmitate ($C_{51}H_{98}O_6$)

$= 51(12.0 \text{ g}) + 98(1.01 \text{ g}) + 6(16.0 \text{ g}) = 807.0 \text{ g/mole}$

c.

activation		$\rightarrow$	-2 ATP
8 acetyl CoA	$\times$ 12 ATP/acetyl CoA	$\rightarrow$	96 ATP
7 $FADH_2$	$\times$ 2 ATP/ $FADH_2$	$\rightarrow$	14 ATP
7 NADH	$\times$ 3 ATP/NADH	$\rightarrow$	21 ATP
		Total	129 ATP

d. $1 \text{ pat butter} \times \dfrac{0.50 \text{ oz butter}}{1 \text{ pat butter}} \times \dfrac{1 \text{ lb}}{16 \text{ oz}} \times \dfrac{454 \text{ g}}{1 \text{ lb}} \times \dfrac{80. \text{ g fat}}{100. \text{ g butter}} \times \dfrac{1 \text{ mole fat}}{807.0 \text{ g}}$

$\times \dfrac{3 \text{ moles palmitic acid}}{1 \text{ mole fat}} \times \dfrac{129 \text{ moles ATP}}{1 \text{ mole palmitic acid}} \times \dfrac{7.3 \text{ kcal}}{1 \text{ mole ATP}} = 40. \text{ kcal (2 SFs)}$

e. $45 \text{ min} \times \dfrac{1 \text{ h}}{60 \text{ min}} \times \dfrac{750 \text{ kcal}}{1 \text{ h}} \times \dfrac{1 \text{ pat butter}}{40. \text{ kcal}} = 14 \text{ pats of butter (2 SFs)}$

CI.39 **a.** The N terminal amino acid is glutamic acid (Glu).

   b. The C terminal amino acid is glycine (Gly).

   c. The nonpolar amino acids are glycine (Gly), leucine (Leu), and proline (Pro). The polar neutral amino acids are serine (Ser) and tyrosine (Tyr).

   d. The acidic amino acid is glutamic acid (Glu) and the basic amino acids are histidine (His) and arginine (Arg). At the pH of the body (about 7.4), glutamic acid will have a negative charge while histidine and arginine will be positively charged.

Glutamic acid (Glu)   Histidine (His)   Arginine (Arg)